MANAGEMENT
SCIENCE
APPLICATIONS
Computing and Systems Analysis

MANAGEMENT SCIENCE APPLICATIONS
Computing and Systems Analysis

Hamed Kamal Eldin

Professor, School of Industrial Engineering and Management
Oklahoma State University, Stillwater, Oklahoma

with

Hooshang M. Beheshti

School of Business and Professional Studies
Radford University, Radford, Virginia

NORTH HOLLAND
New York • Oxford

Elsevier North Holland, Inc.
52 Vanderbilt Avenue, New York, New York 10017

Distributors outside the United States and Canada:

Elsevier Science Publishers B. V.
P. O. Box 211, 1000 A E Amsterdam, The Netherlands

Library of Congress Cataloging in Publication Data

Eldin, Hamed Kamal, 1924–
    Management science applications: computing and systems analysis

    Includes bibliographies and index.
    1. Management—Mathematical models. 2. Operations research.
    I. Beheshti, H., joint author. II. Title.
HD30.25.E42   658.4′032   80-21313
ISBN 0-444-00422-X

Copy Editor   Michael Gnat
Desk Editor   Louise Calabro Schreiber
Design   Edmée Froment
Design Editor   Glen Burris
Art rendered by Vantage Art, Inc.
Production Manager   Joanne Jay
Compositor   Maryland Composition
Printer   Haddon Craftsment

Contents

Preface

The objective of this book is to provide the management decision maker with sound foundation in the application of the many tools and techniques of management science. As a means to accomplish this goal, the book outlines a systematic approach to be followed when identifying and solving management science problems.

The role of the management scientist is to coordinate the efforts of the operations researcher constructing the model, the systems analyst developing the system's design, and the computer specialist providing the needed computing ability. This role, therefore, starts with the choice of an area of application of management science. This is presented in the form of a problem to be solved, and then the model that would represent the problem is developed. After the necessary model verification, the management scientist's function is to underline the logical structure of the problem to identify the tools and techniques necessary to resolve it. Finally, the management scientist translates the results obtained into the form of management decisions, aids their implementation, and performs sensitivity analysis for possible variances.

Management science problems can be identified first by *content* for selection, then by *form* for analysis and solution. Identifying the problem by content leads to the development of the mathematical model of the problem: identifying the relevant variables, formulating the objective function, and constructing the appropriate constraints. Categorizing the problem by form leads to underlining the tools and techniques needed to solve the problem and reduce it to management decisions.

The tools and techniques available to the management scientist can be categorized as mathematical, probability and statistical, computing, modeling, simulation and optimization. Our intention in this book is to give the reader an understanding of where, when, and why the different available tools and techniques needed for the problem solving process apply. As a model for implementation, several applications are outlined related to the areas of research and development, production, marketing, finance and accounting, and industrial relations. As a conclusion, the book evaluates the present state of the art of management science and speculates on the future trends in management science tools, techniques, and applications.

While most books of management sciences concentrate on quantitative methods of operations research and mathematical optimization methods, the emphasis in this book is directed to providing the management scientist with a methodology for applying these tools and techniques to the decision-making process. This book, therefore, is addressed to a wide audience from managerial levels to students of management science. In the universities, this book can be used as a text for senior and graduate level courses of engineering and business administration. It can also serve as a manual and a handbook for the different management levels, as well as for management science practitioners.

The fundamental approach followed in this book will provide the reader with both the knowledge presently necessary and enough background to update this knowledge with regard to advances in the field. The organization of the book progresses in three parts:

PART I. MANAGEMENT SCIENCE TOOLS AND TECHNIQUES

Chapter 1 offers a general introduction to management science, its tools and techniques. Chapters 2–5 present and evaluate different tools and techniques available to the management scientist. They give the reader an understanding of where, when, and why these tools and techniques apply in the management decision-making process. As a summary of the management science process, Chapter 6 develops a comprehensive procedure for the selection of the appropriate management science tools and techniques needed for the solution of management science problems.

PART II. MANAGEMENT SCIENCE APPLICATIONS

This section illustrates how management science tools and techniques are applied in the practice of managerial functional areas. Chapter 7 is devoted to the major systems applications that affect an entire structure of the organization, such as systems in the areas of finance and accounting, personnel, and logistics. Chapter 8 covers minor systems applications that are confined to a single functional part of an organization, such as systems in the areas of research and development and market research.

PART III. MANAGEMENT SCIENCE PROBLEMS AND PROMISES

Chapter 9 gives an evaluation of the present state of management sciences and speculates on its future trends of growth in available tools and techniques as well as applications.

At the end of each chapter, a list of recommended articles is included. These lists represent some good articles written on the application of management science tools and techniques.

Two appendixes related to subject areas of specific interest to management scientists are included at the end of the book. Appendix 1 is a review of available computer software for accounting and financial analysis. Appendix 2 is a list of abstracts for some good articles related to management science tools and techniques to help the reader interested in a specific topic. These articles were chosen as a result of a search of literature published during 1971–July 1979.

Hamed Kamal Eldin

Acknowledgments

The publication of this book in itself will express my gratitude to those who have over the years encouraged me to write it. I have followed their advice in attempting a new approach to looking at the interaction between management science, computing, and systems analysis. The manuscript has benefited a great deal by critical comments and suggestions by many colleagues. My appreciation goes to Max Croft, Sadek Eid, and B. Khoshnevis for reviewing earlier editions.

A special acknowledgement belongs to my co-author Hooshang Beheshti. The book could not have been completed without his considerable contribution.

For proofreading the revised manuscript, I wish to thank my daughter, Amany Eldin. Also, I want to acknowledge my debt to Gwen Starks for her professional contribution in typing the manuscript.

Finally, I wish to acknowledge Elsevier North Holland Publishing Company and particularly Kenneth J. Bowman for his editorial support and Louise Calabro Schreiber for her helpful suggestions. She introduced me to Michael Gnat who spent many hours reading my manuscript for errors and cumbersome language. I am indebted to his literary talents and constructive criticism.

Hamed Kamal Eldin

February 1981

MANAGEMENT
SCIENCE
APPLICATIONS
Computing and Systems Analysis

PART I
MANAGEMENT SCIENCE TOOLS
AND TECHNIQUES

The problems that face managers today have existed in varying degrees since the industrial revolution. Scientific tools and techniques have been applied to resolve these problems for only the past 30 years. Management problem solving and decision making only came to the forefront with the recent emergence of new fields such as operations research, management science, systems analysis, and computer science.

I. DEFINITION OF MANAGEMENT SCIENCE
AND OPERATIONS RESEARCH

There remains a great deal of confusion regarding the meanings of the terms *management science* and *operations research*. Wagner maintains that they are synonymous and are used interchangeably by both the Americans and the British. Furthermore, at Yale University the term *operations analysis* has been substituted for management science. According to Wagner, "Synonyms for the term management science are about as numerous and tenaciously adhered to as dialects of English" [1].

Richmond uses the terms operations research and management science interchangeably, defining the latter as "the application of the operations research approach in the general area of management" [2]. Churchman et al. [3] also equate the two:

> O.R. [operations research] is the application of scientific methods, techniques and tools to problems involving the operations of a system so as to provide those in control of the system with optimum solutions to the problems.

The degree of confusion over the terms has reached such a state that Brinckole [4] asserts "There are other names which may mean more or less the same thing [as O.R.]; systems analysis, operations analysis, management science, quantitative method, and so on." It is therefore necessary to begin by stipulating the definition for each of these terms that will be used in this book.

> *Definition of management science: a process of applying various methods of applied sciences* (i.e., all the available scientific tools and techniques) *to the solution of management problems.* Most large organizations today maintain a department of management science which uses various scientific methods (operations research, systems analysis, etc.) in dealing with their management problems. Computers have enormously facilitated the use of such methods.

> *Definition of operations research: an interdisciplinary science concerned with finding methods for using the applied scientific approach to* (a) *aid in management decisions and* (b) *solve management problems.* Operations research is simply one of the tools used in management science in dealing with management problems.

II. THE RELATIONSHIP BETWEEN MANAGEMENT SCIENCE AND OPERATIONS RESEARCH

In order to grasp the relationship between management science and operations research, a short review of their history is in order.

Operations research was developed in 1942 by physicists and mathematicians in England. Due to the fact that it was initially difficult to understand and because its application to real-life problems required vast calculations, scientists did not accord it much interest.

In 1957 computers came into existence, and with their development made possible the use of various operations research techniques in the resolution of difficult problems. The emergence of systems theory catalyzed the integration of the various techniques of operations research (e.g., industrial engineering and computer sciences) into what has become known today as management science.

In 1964, with the commercial introduction of third-generation computers, management science became a very powerful aid to management in dealing with its problems in planning, organizing, developing, operating, and controlling. It has since served as an important tool in helping the managerial decision-making process and has played an important role in dealing with both basic and nonbasic management functions. As a result, operations research has become just one of several techniques used in management science.

When describing the relationship between management science and operations research, Johnson et al. [5] indicate that

> The systems concept provides a framework for integrated decision making. Within this framework a number of tools of analysis have been developed over the years, including the latest, most sophisticated techniques used in operations research. [Management science places] these tools in perspective as they relate to the science and art of managing.

III. THE MANAGEMENT SCIENTIST

What is needed today, therefore, is a specialist capable of coordinating available tools and techniques for the solution of management problems. This *management scientist* should be able to develop a system methodology, attack problems in their entirety rather than piecemeal, and organize the contributions of outside disciplines essential to the overall requirements of a system output. Therefore, the management scientist must

1. have high analytical ability;
2. know the methodology needed for systems work and be up to date on the technologies used in system improvements and implementation;
3. be familiar with (though not necessarily an expert on) the use of tools and techniques such as queuing theory and information theory in terms of the inputs they can accommodate, the relationships they can handle, and the outputs they can produce;
4. be able to communicate with specialists in these techniques from other disciplines; and
5. understand the real-world situation and how the organization faces the problem.

IV. MANAGEMENT PROBLEMS

The questions facing management arise from the various sectors of the organizational structure and are considered to be problems in *content*. They are directly related to the major departments of a large organization and are classified as problems in research and development, production, marketing, finance, accounting, and industrial relations.

The management scientist, whose task it is to apply scientific knowledge to the solution of management content problems, has ascertained from experience that the majority of these problems are composed of one or more basic structures or forms. These forms, into which the management scientist can classify most management problems, are inventory, allocation, queuing, sequencing, routing, replacement, and competition. A given problem need not be confined to one classification; for example, some inventory problems can be reformulated as either allocation or queuing problems.

After problem classification, a model of the specific situation must be developed. The symbolic or mathematical model is preferred and most frequently used. It allows the component variables of the problem to be related by mathematical and logical symbols and expressions. Aside from presenting complex situations in the easiest and most economical manner, the mathematical model provides quantification of the problem, thereby helping the manager clarify it and reach a solution. To derive a solution from these developed models, the management scientist must apply certain tools and techniques.

A. Techniques

In general, a technique is the application of certain fundamental operations to solve a management problem. The technique consists of basic operations known as tools.

The techniques utilized in formalistic problem solving may be broken down into two types: *deterministic* techniques, which include break-even analysis, improvement curves, critical path method, time series analysis, dimensional analysis, and symbolic logic, and *stochastic* techniques, which include heuristic modeling, stochastic programming, sensitivity analysis, decision theory, competitive modeling, queuing theory, statistical quality control, PERT, Monte Carlo simulation theory, behavioral modeling, Markov processes, and simulation. The difference is that deterministic techniques use data that are assumed to be constant, whereas stochastic techniques apply statistics and probability to give relative answers. In all these techniques, a computer is necessary if an in-depth study is desired.

Of course, there are other modeling, simulation, and optimization techniques that lend themselves to the solution of managers' problems; however, it is the above techniques that are to be considered in this book.

B. Tools

To use the above-mentioned management techniques effectively, certain basic tools are necessary. These tools are not restricted to any one technique. Generally speaking, the deterministic techniques use vast groups of mathematical tools. For example, linear programming utilizes a great deal of matrix algebra. Representing the data as a matrix and using the identity matrix, one can employ the Simplex method to solve linear programming models. Graphical solutions of linear programming models require the use of algebra as well as analytical geometry.

The stochastic techniques rely more on statistics, probability, and distributions than do the deterministic techniques. Sensitivity analysis uses differential calculus to find the minimum point on a cost curve, probability to determine variations from this point, and expected value to evaluate this situation of deviation from the optimum point.

V. DESIRABLE TRAINING FOR MANAGEMENT SCIENTISTS

The question of what constitutes suitable training for a management scientist is not easy to answer. The factors involved are myriad and so broad that a complete delineation would be extremely difficult. *Desirable* background would include far more theoretical training and practical experience than an individual could gain in a lifetime. Minimum requirements, however, can be realistically considered.

As a prelude to training requirements, two broad areas should first be distinguished—systems methodology and systems tools and techniques. Although the line dividing them is at times blurred, tools and techniques would generally include such things as mathematics, probability, statistics, computer technology, modeling, simulation, and optimization. Methodology would involve a study of the systems approach to problem solving, including systems analysis and evaluation.

In the opinion of this author, the management scientist should be an expert in systems methodology and be familiar with the application of tools and techniques to particular situations. Figure 1 shows the technical capabilities in disciplines required for the management scientist.

Course work in methodology should definitely be included in every system-oriented program. This course work should provide a firm foundation in systems management methodology. Its blend of theory coupled with practical examples will furnish a degree of understanding or "feel" that is invaluable to the student. It should succinctly distill the thinking of considerable experience in the field regarding basic definitions.

Logically following are vital facets of methodology: systems function, systems development cycle, systems analysis and design, control system theory, feasibility and trade-off studies, operating procedures and specifications, systems effectiveness and evaluation, reliability engineering, and computers and information systems.

The many tools and techniques applicable in the systems process may be categorized as follows:

Mathematics, Probability, and Statistics. The systems person makes extensive use of probability and mathematical statistics. The theories of probability, distributions, and statistics are used in applications concerned with decision theory and decision making under certainty, risk, and uncertainty. Forecasting techniques owe many achievements to exponential smoothing and mathematical regression.

6

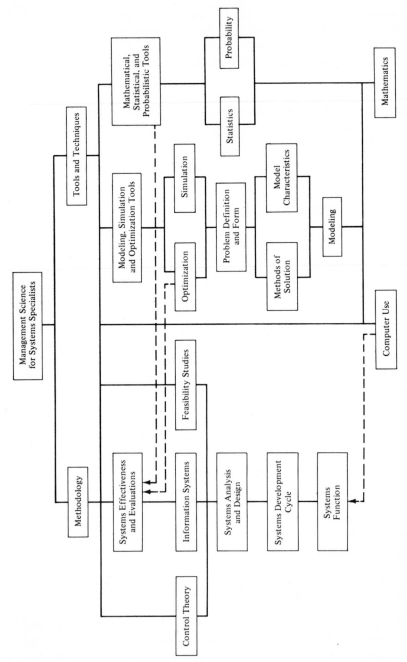

FIGURE 1. Technical capabilities in disciplines required for the management scientist.

Computers Use. Successful application of computers in processing data and supplying results is usually economical and essential in accomplishing the systems effort. Besides being able to understand the potential of the computer and to recognize its applicability, the systems person should also be able to communicate effectively with computer specialists.

Modeling, Simulation, and Optimization. Mathematical modeling and simulation are powerful tools for systems work. The key is to identify the problem in form and content. Methods of solution, whether analytical, enumerative, or deterministic, are secondary to problem definition. Also, characteristics of the mathematical model—whether it is deterministic or stochastic, linear or nonlinear, static or dynamic—should be decided after thorough investigation of the problem form.

Simulation techniques are very powerful in handling queuing problems. PERT and CPM are suitable for sequencing problems. Optimization techniques, such as linear programming and dynamic programming, are useful in handling allocation and routing problems. Value analysis and decision theory are used in selecting from among alternatives in replacement and inventory problems. Figure 2 shows a hierarchy for the suggested areas of study.

A final comment is desirable concerning the need for training in management science. Many of today's management scientists do not have a formal educational background in the desired methodologies and tools

FIGURE 2. Hierarchy for suggested areas of study.

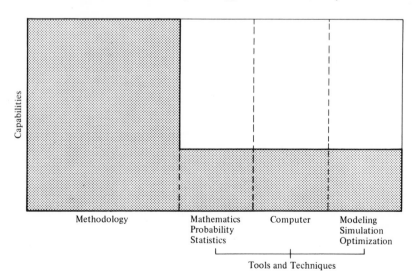

and techniques. However, surveys indicate that the academic institutions should and can provide programs to cover the whole spectrum of systems work so that the next generation of management scientists will be better equipped to meet future challenges.

REFERENCES

1. Wagner, H. M., *Principles of Operations Research*. Englewood Cliffs, New Jersey: Prentice Hall, 1969.
2. Richmond, S. B., *Operations Research for Management Decisions*. New York: Ronald Press, 1968.
3. Churchman, C. W., Ackoff, R. L., and Arnoff, E. L., *Introduction to Operations Research*. New York: Wiley, 1957.
4. Brinckole, W. D., *Managerial Operations Research*. New York: McGraw-Hill, 1969.
5. Johnson, R. A., Kast, F. E., and Rosenzweig, J. E., *The Theory and Management of Systems*. New York: McGraw-Hill, 1973.

Chapter 1
Management Science

This chapter was written with several goals in mind:

1. To give the reader an understanding of how management science can be used in the management decision-making process. Management science is the process of applying scientifically oriented tools and techniques to management problems. These tools and techniques include operations research, systems analysis, computers, and other scientific methods.

2. To describe the evolution of operations research during World War II and its limitations in the postwar era. The systems approach and the development of the electronic computer in the mid-1950s made possible solutions to operations research problems. Management science emerged as a recognized profession and presently continues to exploit the best technology available for the benefit of management.

3. To describe the basic management science tools and techniques. The systems analysis approach and cycle are explained, classic operations research applications are explored, and computing tools and their application in the management process are analyzed.

4. To explain the distinction between the problem by content, as viewed by the manager, and the same problem by form, as visualized by the management scientist. Methods of finding solutions to problems by form are briefly outlined.

INTRODUCTION

Management has traditionally been considered an art learned through experience or handed down from generation to generation within family-owned corporations. However, early in the 20th century, beginning with the work of Frederick Taylor [1], scientific principles were employed in the field. There has gradually emerged a professional class of managers who by means of formal training and education can observe and interpret experiences and apply scientific procedures to management. Thus, management has been in a process of transition for some years and could best be characterized at present as consisting of both art and science.

Management

There are about as many definitions of management as there are writers in the field. Beech [2] has defined management as the process of utilizing material and human resources to accomplish designated objectives. Management goals are met by the proper organization, direction, coordination, and evaluation of people. Newman and Summer [3] have stated that the total task of management can be divided into organization, planning, leading, and controlling. Chester Barnard [4] has observed that the essence of management, at least its executive function, is the exercise of effective leadership and decision making. Wakefield [5] has stipulated that management consists of four basic functions: research and development, production, marketing, and finance and control. Three additional activity areas that influence the work environment are personnel management, external relations, and secretarial and legal affairs. Wakefield further cites 11 functions of managing that are common to all levels of management:

1. Information gathering
2. Information synthesis
3. Planning
4. Decision making
5. Organization
6. Communication
7. Motivation
8. Direction, guidance, and counseling
9. Measurement, evaluation, and control
10. Development of people
11. Promotion of innovation

We must also note that management is an activity involved with accomplishing results through the efforts of other people and that the most difficult problems managers face are usually human rather than technical.

It is against this background of difficult and diverse factors that management has been evolving.

Management Science

Just what is a science, and how does it relate to management, and to so-called management science? The word science comes from the Latin word *scientia,* meaning "a knowing." Science has been defined as a specific accumulation of knowledge systemized and formulated with reference to the discovery of general truths or the operation of general laws. The scientific method is usually employed: an assumption (*hypothesis*) is made regarding the existence of some general truth or law and an experiment is then performed to yield measured data by which the hypothesis may be accepted or rejected. Thus we may recognize whether a certain body of knowledge has become a science by certain indications. First, the significant parameters must be measured, and only those that result in the same measurement made many times by many observers are judged to be facts. Hypotheses must be formed that account for the behavior of known facts, and these hypotheses tested in every conceivable manner. A hypothesis that has not been disproved over years of testing will ultimately be considered a law. Finally, theories devised and tested in a similar manner will account for the behavior of the laws. Thus, when sufficient theories and laws have been accumulated for a specific body of knowledge such that the general truths and the operation of the general laws are known, then we may consider that body of knowledge to be a science. The steps of scientific procedure can be summarized as follows:

1. Formulation of the problem
2. Construction of a model to represent the system under study
3. Derivation of a solution from the model
4. Testing of the model and the derived solution
5. Establishment of controls over the solution
6. Implementation of the solution

There have long existed various management tools, techniques, and laws for solving isolated problems. What has been needed is a structured science that tries to solve management problems in a systematic way. This new science is referred to as *management science* and, as in any new field, a common terminology has not yet developed. However, in this book management science is defined as the *application of the scientific method to the solution of management problems.* It is multidisciplinary, encompassing all science-based bodies of knowledge required in the solution of any given problem.

The aim of this new discipline is to determine the most favorable course of action from all the available alternatives for any given set of management circumstances. The basic method is to study a problem by considering the total system (all significant factors, both human and material) and to analyze the resulting (hopefully mathematical) model by means of operations research techniques. Thus the basic tools of management science are systems analysis and operations research. Since lengthy and complicated calculations are often necessary, the modern digital computer is also usually employed.

I. EVOLUTION OF MANAGEMENT SCIENCE

Although the physical sciences evolved several hundred years ago, it has only been comparatively recently that progress has been made in the social and behavioral sciences, including management science. Prior to the industrial revolution, most industrial enterprises were very small in terms of employees, space, and output. Their operating and technological problems were much smaller than those of modern organizations; they were usually owned and managed by the same individual. However, with the coming of the industrial revolution, enterprises expanded, technology became complex, production was mechanized, labor became specialized as to function, and it was no longer possible for a single individual to perform all the management functions. Furthermore, as industries grew, companies merged or expanded into multiplant operations, thereby greatly expanding the management function. This created a demand for a professional class of salaried managers which emerged over the years, soon dominating management and contributing largely to the advancement of the field.

In the early 1900s Frederick Taylor, the acknowledged father of scientific management and its initial driving force, introduced the principle of experimentation and the method of developing a scientific procedure for each element of a particular job. He created management techniques such as methods study, standardization of tools, differential piece rate systems, and cost control systems. He also promoted the concept of cooperation between management and labor as opposed to the widespread autocratic attitudes on the issue.

Taylor prescribed the duties of managers as follows:

1. To develop a science for each element of work to replace old "rule of thumb" methods
2. To select, train, teach, and develop the worker scientifically
3. To cooperate with labor to ensure that all work is done in accordance with the principles of the science that has been developed

4. To divide both work and the responsibility between management and labor on the basis of who is best qualified for the given task

Taylor's program and philosophy can be summarized in four concepts:

1. Science, not rule of thumb
2. Harmony, not discord
3. Cooperation, not individualism
4. Maximum, not restricted, output.

Taylor's ideas led to great progress in the field of management. However, his teachings were based on the concept of the "economic man," motivated to maximize individual gain. His approach ignored sociological and psychological factors inherent in humankind, and further advances in management had to await developments in the social and behavioral sciences.

An extensive research program conducted at the Chicago-based Hawthorne Plant of the Western Electric Company during 1927–1955 led to great advances in the field of human relations. This program was led by Elton Mayo, who is now considered to be the father of industrial sociology. The program consisted of experiments on illumination, relay assembly test room studies, a mass interview program, bank wiring observation, and personnel counseling.

The results of the Hawthorne Studies are summarized in some detail by Mayo [6], Roethlisberger and Dickson [7], Smith [8], Whitehead [9], and Dickson and Roethlisberger [10]. These studies originally sought to determine the effect of environmental factors such as light intensity, length of work day, and coffee breaks on productivity. The experiments failed to find a direct relationship between these factors and productivity; however, they did ascertain that sociological and psychological factors can influence productivity more than physical factors. The well-known "Hawthorne effect" was also studied in relation to the effect of management attitudes on worker productivity. It was also observed that workers formed social groups that exercised effective control over the behavior of individual members and their output. Output was frequently restricted due to social group pressure even though it meant lower total wages for individuals. Thus it was determined that the worker is a social creature who is not entirely materialistic, a fact that must be taken into consideration by a successful manager.

The Hawthorne studies represented the beginning of the human relations movement and were vital to continued progress in management science. In 1938 Chester Barnard qualitatively incorporated these results into a workable theory of management that is still studied by practicing managers.

The state of knowledge regarding human relations has continued to advance since the Hawthorne studies. However, as noted by Beech [2], no widely accepted general theories have been developed, and this science is still only in its infancy.

As noted by Smiddy and Naum [11], it was during World War II, spurred by the fires of necessity, that the evolution of scientific management accelerated. The immensity of global war and its resulting logistics problems, the new technology, the overwhelming demands on the free enterprise system, as well as the critically felt need for survival speeded the acceptance and advancement of management science theories. According to Ackoff and Rivett [12], the successful introduction and development of operations research occurred at this time as a response to the need to handle problems of such overwhelming magnitude.

A. Evolution of Operations Research

In 1939 at the outbreak of World War II in Europe there was a small British operations research group in existence. This group was studying the problem of integrating the newly developed radar system of early warning against enemy air attack with the older system which was based primarily on visual sighting and identification of planes. The results of their study were the identification of certain weaknesses in the system and the recommendations for improving operation techniques. The group's work was highly successful. It has been estimated that the introduction of the radar system increased the probability of intercepting enemy aircraft by a factor of ten. In addition, the operations research group's effort increased the probability by a further factor of two.

The mixed-team approach was, from the start, an important feature of an operations research group. The initial group consisted of people with physical science, mathematics, and statistics backgrounds. Another unique feature of the World War II operations research group, and probably its most clearly distinguishing feature, was its consideration of the problem in terms of its relationship to the entire operation, thereby necessitating the study of additional operationally related problems. Today, this consideration of the problem in terms of its relationship to the entire operation is commonly referred to as the systems approach (i.e., the total systems approach).

Since World War II progress in operations research has continued to be rapid. Those participating in operations research during the war tended to move to industry after it was over, and continued to develop techniques within the field. It has been noted [13] that in the past 30 years four major developments have stimulated the growth of management science:

1. Feedback control theory
2. A better understanding of the decision-making process

3. An experimental approach to business system analysis
4. The digital computer.

A feedback system utilizes the principle of comparing the present environmental state to a previously established desired condition. A decision is made based on relevant data, which results in actions affecting the environment and thus influencing further decisions. If the system is composed of electromechanical devices, it is termed a *servomechanism*. Such systems are the basis of automation, which has had such an impact on industrial production and on other aspects of management science. If the system is composed of people and the feedback is communication among them, then it is termed the *decision-making process*.

The experimental modeling of business systems via operations research is a powerful means of studying the behavior of a system and is greatly facilitated by the digital computer. Such computers provide a sort of management laboratory within which experiments and evaluations of complex business situations and decisions can be made before they are put into effect.

B. Management Science as a Profession

In 1953 a professional society, the Institute of Management Sciences, was formed by interested engineers, scientists, and managers to extend the body of knowledge in the field and to promote the scientific understanding and practice of management. Another society was formed at about this time under the title of "operations research," whose objective was the improvement of complex systems of all kinds. Churchman [14] has noted that the cohesive force in operations research has been the development of the discipline and profession, whereas in management science it has been a dedication to the improvement of the human environment and thereby of the human being.

C. Management Science vs Operations Research

It seems pertinent at this point to compare management science and operations research to determine the extent of their similarities and differences. Wagner [15] and Trammell [16] have noted that many authors equate management science with operations research or use one term as a substitute for the other. Within this book, *operations research* is defined as the interdisciplinary science concerned with *using quantitative techniques to assist management* in solving problems and making decisions. This discipline is generally concerned with mathematical tools and scientific and engineering methods for studying physical processes and does not attempt to treat social and behavioral problems. Management science, on the other hand, is the process of applying the methods of applied science to the solution of management problems and is therefore concerned with applying operations research results and the results of other science-based bodies of knowledge (including the social and behavioral sciences). In other words, operations research is primarily the concern of the mathematician, scientist, and engineer, whereas management science is management-oriented, directed toward the application of operations research (and other scientific) results to the management decision-making process.

It is obvious, then, that the two disciplines overlap in their application of procedures and scientifically oriented tools; yet each retains an independent identity of its own. In any event, both management science and operations research are simply a means to an end. They seek to place before management the best possible alternatives to given problems as a means of aiding and improving management decisions. In the end, it must be management that makes the final decisions—as always, by means of inevitable value judgments. For the purpose of discussing the organizational location of the management science function, the main tools and techniques of management science will be considered separately. As already noted there is considerable overlap.

II. OPERATIONS RESEARCH

We have defined operations research as the interdisciplinary science concerned with using quantitative techniques to assist management in solving problems and making decisions. We noted that operations research is generally concerned with mathematical tools and the implementation of scientific and engineering methods in the study of physical processes. We further emphasized the independent identity of management science and operations research in spite of obvious overlap. Symonds [17] defines operations research as the problem-solving objective and management science as the development of general scientific knowledge, maintaining that the two are complementary.

It is beyond the scope of this book to explore the mathematical details and techniques of the various operations research tools. These are well covered in textbooks by Wagner [15], McMillan and Gonzales [18], and Teichroew [19]. Rather, we shall address ourselves to the main concepts and characteristics of operations research in order to better understand the management science process.

A. Characteristics of Operations Research

A brief survey of the relevant literature reveals disagreement over just what operations research really is. The following definitions of operations research are the most common.

What operations researchers do

The science of decision

Quantitative common sense

The art of giving bad answers to problems to which otherwise worse answers would be given

A simultaneous application of industrial engineering, quality control, civil engineering, statistics, market analysis, applied math, applied physics, etc.

Concerned with conducting studies to help decision makers make better decisions

The application of scientific methods, techniques, and tools to problems involving the operations of a system so as to provide those in control of the system with optimum solutions to the problem

An aid for the executive in decision making because it provides the needed quantitative information collected on a scientific basis

Applied science in the best tradition of modern engineering

These definitions individually may lack precision but they do provide for a listing of the main characteristics of operations research:

1. Total systems orientation
2. Application of the scientific method
3. Development of a model
4. Use of special mathematical tools and techniques

The use of the scientific method in total systems evaluation is viewed by some as the essence of operations research, the measure by which one can determine whether a given technique is in fact operations research or something else. Others believe that the crucial criterion is the skillful use of special mathematical techniques and models. We shall assume that anything satisfying the four characteristics listed above is operations research. Perhaps the one characteristic of operations research that distinguishes it from more familiar sciences such as physics, chemistry, and mathematics is the fact that its explanation of phenomena asserts little about the exact physical nature of the elements involved, but rather deals with their interplay.

B. Phases of the Operations Research Process

Laue [20] has observed that in the execution of any given operations research project we may distinguish the following seven phases:

1. Problem formulation
2. Development of model
3. Data collection

4. Analytical solution of model
5. Testing of the solution
6. Implementation
7. Monitoring of the new system

Perhaps the most difficult and important part of any study is the formulation of the problem statement. The definition of the problem, the determination of all variables that significantly influence the problem, and the selection of the results to be obtained are all critical to the success of the study. Furthermore, if the results of an operations research study are to be worthwhile, they must contribute new information about the problem, in a useful form, to help management make a more intelligent decision.

The second phase requires the development of a realistic model. In terms of operations research the model is simply a representation of the subject of inquiry. Its purpose is to reduce the situation under study to a frame of reference that permits convenient manipulation of the pertinent variables. Such models will usually be in either analogue or symbolic form. Analogue models use one physical property to represent the characteristics of another. For example, contour lines on a map represent actual terrain and voltage magnitude represents force or velocity on an analogue computer. In symbolic models components and their interrelationships are assigned symbols, as in an algebraic relationship. Symbolic and analogue models are more abstract and harder to visualize than other models, but can represent system parameters more accurately.

Care must be exercised during the development of the model to ensure that

the model is capable of yielding results that represent a true measure of the system;

all significant parameters or inputs that influence the system, as well as their true interrelationships, are included;

the model indicates the correct procedure for adjusting the control variables in order to maximize effectiveness, minimize cost, and so on.

Data collection can be a problem in an operations research study. Obtaining, selecting, and processing data relevant to the study as well as testing for data validity are important functions. However, actual results are not always easy or possible to obtain and the techniques of simulation are often used instead.

The fourth phase comprises the analytical solution of the model. The original model may be solved mathematically in closed form, the accumulated data may be used to arrive at a numerical solution, or a simulation study may be performed on the model. Simplifying assumptions will have to be made in any of these cases and careful judgment used if meaningful results are to be obtained. Thus the assumptions made for the study should be fully justified, and the mathematical model chosen should minimize the need for assumptions.

After obtaining a solution it is necessary to test its validity before recommending implementation. These tests involve the examination of the sensitivity of the solution to changes in the assumptions and environment of the system. Such examination can point out where the refinement of input data is most needed, indicate the affects of system changes on parameter values and how these changes in value can be evaluated, and render the results of an operations research program more amenable to management's inituitive judgment.

Those recommendations that management decides to put into operation should be implemented under the guidance of all members of the *systems team*, which of course includes operations research analysts. If the results of the operations research study are implemented, we can be assured of the successful role of operations research in the decision making. The continued implementation of the results of operations research studies will attest to operations research as a worthwhile input to the decision-making process.

Monitoring the new system is the last phase of the process. Such monitoring must also be carried out by the entire systems team to ensure the correct utilization of the system and to evaluate any possible condition changes following implementation. Continued monitoring allows for modification of the system if conditions do change.

C. Solution Techniques

We have already noted that managers like to consider problems in terms of their content. This is because each problem is thought to be different from those confronting other managers. Although it is true that few problems ever have exactly the same content, most management problems can be reduced to one of several common forms. Operations research must address itself to the forms of problems. With this in mind, we may divide the management science process into the following areas:

1. Problems in content—management
2. Problems in form—operations research
3. Analysis and solution—management science

Any mathematical technique can be a useful tool in operations research. Normally we divide such techniques into deterministic and stochastic (Part I, Section I.B.5), with the nature of the system being studied dictating the type used. Systems that are represented by deterministic models contain no uncertainty, and changes in conditions can be predicted. In such cases the methods most commonly used include matrix algebra, differential and integral calculus, transcendental functions, differential equations, and some of the so-called special functions.

However, since many if not most systems cannot be defined without some degree of uncertainty, we must also resort to probabilistic models and the stochastic process. In such cases we find that results yield probable rather than exact solutions. Obviously the principle methods employed are derived from probability theory and mathematical statistics, including techniques such as point and interval estimation, testing of hypothesis, exponential smoothing, linear and multiple regression, correlation, and analysis of variance. The tools and techniques used for management science work are discussed in more detail in Part II.

Before we attempt to relate management science to the management systems function, one other aspect of the systems approach or scientific method will be mentioned—that represented by the term *systems analysis*. This concept is vital to the investigation of the characteristics of any system and will be shown to be an essential phase of the design and development of a management system.

McMillan and Gonzalez [18], frankly state in their book *Systems Analysis* that "the authors are not certain what is meant by systems analysis." Johnson et al. [21] state that systems analysis includes the process of establishing objectives as well as the evaluation of alternative proposals and that it must involve cost-effectiveness trade-offs. In our definition of operations research we noted that it was mainly concerned with mathematical tools and scientific or engineering methods and generally did not attempt to treat social and behavioral problems. However, systems analysis can in principle treat total socioeconomic systems. According to McMillan and Gonzalez [18] successful experimental validation of this fact is not yet conclusive.

We shall define systems analysis as the *systematic approach to the study of the behavior of systems and to the establishment of systems objectives and alternatives with regard to cost trade-offs,* which provides executives with a sound quantitative basis for decision making. Systems analysis makes use of other tools of management science, such as operations research and computers, to carry out quantitative procedures. Furthermore, it emphatically requires that consideration be given to the total system and that decision making be based both on quantitative results and the personal value judgments of managers.

III. THE SYSTEMS FUNCTION

This section describes the function of designing, developing, installing, and evaluating systems that provide information for management. It includes a brief look at the history of the systems function, its primary responsibilities, and its location in the firm's organizational pattern. In Section V, the relationship of the systems function to the management science organization is established.

New scientific management techniques and the rapid infiltration of electronic data processing into a wide range of management activities have been responsible for the recent rise and widespread growth of the systems function. Systems work is geared to the increasing complexity of successful business management. Management research has bypassed the former limits of the systems and procedures analysis and audit function and has developed techniques that make the design of broad-scope management systems both possible and practical. These systems, especially those capable of handling the increasingly complex business projects, are based on the principles of scientific management techniques such as operations research and analysis, statistics, value analysis, and decision theory, *and* a computing tool powerful enough to carry out these techniques.

Every company, large or small, has and always has had information systems of one kind or another. The person in charge of systems and procedures has traditionally had to ensure that reports, records, and data were collected systematically and on time, to be presented to the proper authorities for analysis and action. Until a few years ago, the information or "management" system was associated with the comptroller function. When electronic data processing appeared on the scene, it was restricted almost completely to payroll systems, cost accounting systems, and the like. The data processing function, including both equipment operation and systems application, was therefore placed under the comptroller.

As data processing capabilities expanded, the application of computer methods naturally became more varied. More and more systems of different origins were created, many of which required electronic data processing. It soon became apparent that management systems were required in all phases of company activity and that the need for electronic data processing was just as widespread.

As larger computer complexes became available for individual companies, an increased need arose for centralized control of computer resources. This centralized control was often placed in a data processing division usually reporting to the comptroller.

The need for centralized control of the systems effort also increased. Although systems and the systems approach had been present at all levels of management activity for years, more sophisticated, formalized systems were developed to assist in controlling inventory, production, scheduling, and a host of other activities. The systems function slowly evolved as a centralized organizational activity not confined to any single organizational element because of its company-wide application and objectives.

The specialized knowledge required to perform the systems function, and the influence of the earlier comptroller relationship, pushed it into the administrative staff level.

Systems work has thus developed as professional staff work involving many varied responsibilities. Those who perform this work are relied upon for their specialized knowledge and training, the breadth of their views, and their ability to recognize and create valuable innovations. They incorporate not only specialized analytical ability and training in data handling methods, but also knowledge of the systems function itself. This is particularly true of the systems managers at higher corporate levels: as members of management or as advisors to top management, they represent the systems function within administrative councils.

The systems function requires authority that is not at present formally designated to it because of its staff position; however, with time it will establish adequate ideological authority because of its specialized knowledge. One of the principal interim systems responsibilities is to encourage operating managers in systems matters. The systems manager (director of the systems function) can influence operating managers to secure an understanding of the advantages of systems improvement. Beyond this, the systems manager can introduce an understanding of the significance of systems work at the administrative level so that top executives will in turn reflect the values of the systems function to operating managers.

With adequate appreciation from management, the systems function will ultimately recognize its basic objective: to assist management, at all levels, to perform its function more effectively and efficiently. In some situations this means providing information in a more timely and presentable fashion; in others, providing tools for the planning, distributing, and tracking of resources; in still others, providing systematic methods and procedures, standard forms and reports. These situations are multiplying by leaps and bounds. Systems applications are no longer restricted to business administration, although this is a large part of the systems function grouped under "business" or "commercial" systems and includes payroll, accounting systems, and so on. Many applications now involve the management of scientific, engineering, and manufacturing activity. These are grouped under "scientific" or "technical" systems and include linear programming applications, simulation, scheduling and planning systems, and the like.

The staff professional performing in the systems function can have any of a number of titles, some of which are more articulate than others—confusion exists here as with any nomenclature that involves the word "systems." As indicated earlier, we choose to identify people employed in the systems function as *systems specialists* and the person who heads an organization's systems function, the head of the staff group responsible for this function, as the *systems manager*. Those who assist various organizational elements with specific systems matters and who report to the systems manager are called *systems consultants*. Large management systems organizations also employ *systems analysts*. They are usually involved with less sophisticated systems but can translate the design requirements of any system into terms that permit a computer programmer to prepare instructions for processing the system by computer methods. In some organizations, systems analysts are assigned to the data processing division rather than associated directly with the systems function.

Let us differentiate the responsibilities of the systems function from those of the data processing function. The latter usually consists of at least two subfunctions (see Figure 1):

1. Planning
2. Operations

The planning function is mainly performed by computer systems analysts and systems programmers. It is principally the computer systems analyst among all members of the management science team who is responsible for planning.

The computer systems analyst coordinates all aspects of a given project within the data processing organization and initiates action to achieve the following:

1. Review objectives
2. Analyze the present computer system
3. Design a new computer system
4. Provide documentation
5. Prepare test data and supervise a program test
6. Write a final report

The computer systems analyst relays the necessary information to the computer systems programmer, who prepares the problem for actual computer evaluation. The computer systems programmer carries out functions such as:

1. Drawing the block diagram
2. Coding the program
3. Testing the program
4. Debugging the program
5. Carrying out a final test of the program
6. Documenting the system

When the program is ready it is transferred to the operations section where an operator loads the program on the computer and initiates the computer run. The operator schedules the various programs that are run on the computer and supervises the control panel itself. In addition to executing the actual programs on the computer, the operations section

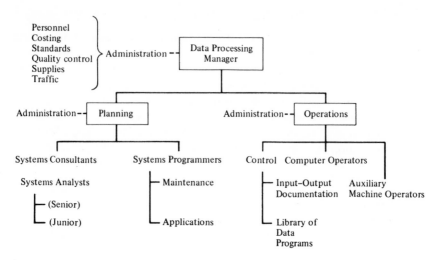

FIGURE 1. Computer systems: data processing organization.

also:

1. Maintains a library of programs and subroutines
2. Maintains a magnetic tape library
3. Processes data into and out of the facility
4. Installs recording media in input–output devices
5. Monitors the overall operation and provides system maintenance as needed

A new staff specialist function has been created as a result of the development of the operating system in third-generation computers. This specialist, called the *systems programmer,* sees that the computer system is designed efficiently and provides assistance to the following:

1. The systems analyst who identifies the procedural problem areas and designs the new system solution
2. The programmer who translates these systems designs into computer programs
3. The operator who runs these programs on a scheduled basis

It is evident that the systems programmer, being most concerned with the entire picture, requires more experience and qualifications than do the other members of the management science team. The systems programmer coordinates the efforts of all groups to see that they function as efficiently as possible and accomplish the most meaningful management objectives.

The following is a list of the systems programmer's responsibilities:

1. Standards and conventions of programming, testing and debugging, operations, documentation
2. Language selection
3. JCL preparation and checkout
4. Macro writing: "tailoring" vendor programs
5. Systems generators
6. Software and hardware selection and "tuning"
7. Application review
8. Personnel training and education

This chapter has not been all inclusive and the reader is directed to more thorough treatments by Desmonde [22] and Thesen [23], and elsewhere [24, 25] for more complete programming information.

As is evident from the above discussion, the task of data processing personnel with respect to management systems is to design *computer* systems and programs to support their operation. The data processing function is only a part of the overall design and implementation of an information system. Other phases of the development cycle of a management system will be described in Chapter 4. Let us contrast the general responsibilities of the systems function with those of the data processing function. There are some elements common to all systems projects:

1. Analysis
2. Problem solving
3. Design (or modification)
4. Pilot testing
5. Implementation
6. Training

The responsibility for these different phases of systems work must be shared by the systems specialists and the operating personnel (the "users" of the system). This sharing of responsibility must persist throughout the planning, design, testing, and installation phases of systems effort if success is to be achieved. This arrangement will help the systems manager develop an effective approach to systems work within the company and determine the best use of the company's resources.

The optimal sharing of systems responsibilities is not always implemented. Table 1 shows the various approaches to responsibility designation. Each of these approaches has its advantages and disadvantages. As indicated, the activities or responsibilities can be divided into planning and installation phases. Planning includes analysis of the problem or situation, discovering the solution to the problem, designing (or modifying) a system to implement the solution, and pilot testing the system to prove

TABLE 1. Possible Approaches to Systems Responsibilities

Systems approach	Planning by	Installation by
A	Systems consultant	Systems consultants
B	Operating personnel	Operating personnel
C	Systems consultant	Operating personnel
D	Systems consultant and operating personnel	Systems consultant and operating personnel

its workability and probable value prior to installation. In many cases this work must be performed outside the loop of regular business activity. That is, the system must be designed and tested before it can become a full-fledged part of the management cycle. The installation phase is therefore the transition from design and testing to implementation as an operating system.

Consider first systems approach A. In this approach, the planning and installation phases are both assumed to be the responsibility of the systems consultant. Operating personnel are not directly involved in the development, testing, or installation of the system; they are only required to maintain the system or furnish data for its maintenance once it is operable.

The disadvantages of this approach are more obvious than the advantages. A glaring disadvantage is the information barrier that exists between the two groups. Operating personnel are far better informed of the work situation and the nature of the problem. The systems specialist can only reach a valid analysis of the problem with the cooperation of the operating personnel. Information transfer is essential in other phases of the systems project, and if such transfers do not occur, resistance to the system will develop among the operating personnel.

Assuming that operating personnel agree to operate and utilize the new or modified system once it is installed, it is practically impossible for them to have an adequate appreciation of the philosophy and general objectives of the system since they have not been involved in its development. At the very least, an appropriate course of instruction will be required during the installation of the system. This assumes, of course, that the operating people will eventually be involved in implementing the system. Only in this case can management have confidence that the results of the system will reflect the true situation.

Systems approach B requires that both planning and installation be the responsibility of the operating personnel. Here, as in systems approach A, the information barrier presents problems. The operating personnel lack knowledge of systems tools and techniques. The systems consultant can hardly recommend appropriate tools and techniques unless familiar with the problem and situation to which they are to be applied. It is also clear that operating departments have more limited, quite different objectives from those of higher-level management. A system designed to meet only departmental needs with no consideration for the needs of others could never furnish the information desired at higher levels.

Systems approach C assigns the planning responsibility to the systems consultant and the installation responsibility to the operating personnel. Although at first glance this approach seems very functional, closer study reveals that problems still exist and the information exchange essential during both phases is restricted.

The most successful approach to the systems function is systems approach D in which the systems consultant and operating personnel share responsibilities in both the planning and installation phases—a joint effort from initial analysis of the problem to completion of systems installation. The systems specialist is professionally equipped with the concepts of the systems function, is aware of the need for a systems approach that satisfies management requirements at all levels, and is an important source of ideas because of past experience and research.

The operator, fully acquainted with the operating situation, is a principle source of operating facts whose early assistance and interest in the development of the system is required for successful implementation. In fact, it is ideally from the operator that the need for the system should arise and that the request for the planning and installing assistance of the systems consultant be made. This ideal arrangement is a direct function of

1. the proper emphasis on the education of operating personnel by top management, including the systems manager, regarding the systems function, and
2. the demonstrated and proven specialized knowledge of systems function personnel.

There is a strong implication in this discussion which should be clarified. In many cases, a systems consultant in collaboration with the appropriate person or persons from an organization's operating element is all that is required to plan and install a system. In some situations where the system requirements are relatively straightforward, a systems analyst can handle the problem alone. However, in the design of more complex systems, a team of specialists should be involved. Again ideally, the operating manager who has detected the need for a system and requested assistance in its development should head the team charged with this responsibility. The systems consultant can, with proper blessing, head the team if the operations manager cannot relinquish time for this task. This should not, however, minimize the importance of the latter's participation.

In addition to the operating manager and the systems consultant, an operations research specialist may be required to assist in the development of the system model or in the simulation of the system for testing purposes. A technical specialist, such as a quality control expert, can be valuable for certain projects. A behavioral scientist can assist in the evaluation of the human aspect, that is, the effect of the system on those persons involved in its operation and use. A systems analyst who is active in the conceptual stages of a large systems program is able to convert the system requirements into computer terms in a much more organized, efficient way than if handed a system specification much later in the planning stage. This team of specialists, decided upon early in the planning phase by the systems consultant and the operating manager, although not required for all projects, is the ideal composition for broader scope systems projects.

In addition to the proper division of responsibility between the systems consultant and the operating personnel is the former's overall expectation of top management:

A clear statement of problems and needs

A clear and objective statement of project priorities

A reasonable time to accomplish staff work

Decisions when required

Corrective action by line function

Suitable line support and availability

Objective appraisal of staff recommendations

Within this framework, top management has every right to expect the systems function to provide professional assistance to management at all levels through

Persistent examination of company functions and activities

Advanced methods of analysis and problem solving

Improved operating information

Better use of corporate resources

IV. MANAGEMENT SCIENCE ORGANIZATION

The use of the electronic computer has had a profound effect on industry in the past few years, but applying the tool to management needs has probably led to the increased employment, and even increased value, of management systems. It has been said that if factory automation or process control represents an extension of the human hand, then the computer in business is an extension of the mind.

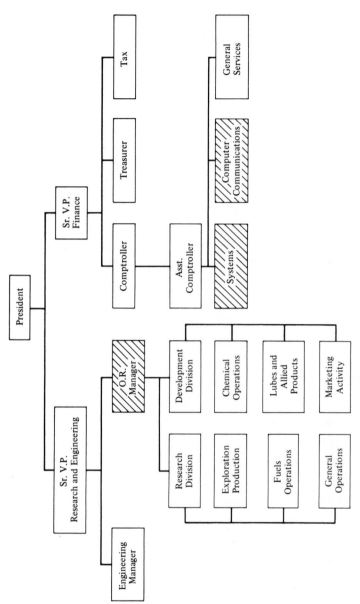

FIGURE 2. Former typical placement of systems analysis and operations research functions.

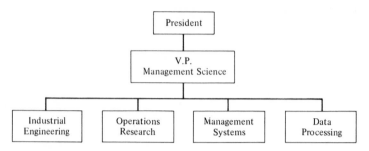

FIGURE 3. Management science organizational chart.

Educating management to view the computer as an extension of the mind and not of the hand is a crucial problem. However, with specific reference to management science, another such problem is the development of professionals who can recognize the capabilities of the latest electronic equipment and of the advanced system approach techniques and then translate these capabilities into tools that a company can use to manage its resources better. This requires a creative, imaginative, highly skilled individual who has mastered the profession and can communicate the capabilities, limitations, and usefulness of these systems to management.

According to Wagner [16] the placement of the management science function within an organization is no longer the subject of much debate among those in the field. In the past, the systems analysis function and the electronic data processing responsibility were often placed within the comptroller's organization because of their involvement with records, reports, company data, and so on. Frequently the operations research function was placed within the research and development of engineering organizations because of its heavy emphasis on scientific methods and techniques (Figure 2). However, with the coming of large centralized computers available to organizations throughout the company, the integration of these functions into a management science discipline, and the emphasis of the total systems approach, we now find the management systems, operations research, and computer functions beginning to be grouped within a management science staff organization reporting to the vice presidential level (see Figure 3). In most cases, practical considerations dictate the location of the function. Thus the present trend is for the management science function to be situated within a company as a high-level staff organization, placed within easy access of top management as well as of all line organizations.

REVIEW QUESTIONS

1. Contrast management and management science.
2. Describe the various steps involved in a scientific procedure.
3. List and explain three major developments that have contributed to the growth of management science.
4. Contrast management science and operations research.
5. Operations research has several distinctive characteristics. Discuss four of them.
6. Discuss seven phases of the operations research process.
7. Define the systems function and show how it can provide information for management.
8. Give three examples of managerial problems that could be solved by using management science techniques.
9. What are the primary functions of a systems programmer?
10. Discuss the different systems approaches that are discussed in this chapter. Which one is the best and why?
11. What are the major responsibilities of a systems consultant?

REFERENCES

1. Taylor, Frederick N., *The Principles of Scientific Management*. New York: Harper, 1911.
2. Beech, Dale S., *Personnel: The Management of People at Work*. New York: Macmillan, 1965.
3. Newman, William H., and Summer, Charles E., *The Process of Management*. Englewood Cliffs, New Jersey: Prentice-Hall, 1961.
4. Barnard, Chester I., *The Functions of the Executive*. Cambridge, Massachusetts: Harvard University Press, 1938.
5. Wakefield, John E., *Some Practical Aspects of Modern Management*. New York: Barrington Assoc., 1959.
6. Mayo, Elton, *Human Problems of an Industrial Civilization*. New York: Macmillan, 1933.
7. Roethlisberger, F. J., and Dickson, W. J., *Management and the Worker*. Cambridge, Massachusetts: Harvard University Press, 1939.
8. Smith, Henry Clay, *Psychology of Industrial Behavior*. New York: McGraw-Hill, 1955.
9. Whitehead, T. N., *Leadership in a Free Society*. Cambridge, Massachusetts: Harvard University Press, 1936.
10, Dickson, W. J., and Roethlisberger, F. J., *Counseling in an Organization*. Boston, Massachusetts: Graduate School of Business Administration, Harvard University, 1966.

11. Smiddy, H. S., and Naum, L., Evolution of a "Science of Managing" in America, *Management Science* 1 (1) (October 1954).
12. Ackoff, R. L., and Rivett, P., *A Managers Guide to Operations Research.* New York: Wiley, 1963.
13. *Introduction to Management Science.* Poughkeepsie, New York: IBM, Code-U950021.
14. Churchman, C. W., *Executive Readings in Management Science.* New York: Macmillan, 1965.
15. Wagner, H. M., *Principles of Operations Research.* Englewood Cliffs, New Jersey: Prentice-Hall, 1969.
16. Trammell, W. D., Management Science: Solutions Looking for a Problem, *The Oil and Gas Journal* (24 March 1969).
17. Symonds, Gifford H., The Institute of Management Science: Progress Report, *Management Science* 3 (2) (January 1956).
18. McMillan, C., and Gonzalez, R. F., *Systems Analysis.* Homewood, Illinois: Richard D. Irwin, 1965.
19. Teichroew, Daniel, *An Introduction to Management Science.* New York: Wiley, 1964.
20. Laue, Hans L., Operations Research as a Tool for Decision Making, *Journal of Industrial Engineering* XVIII (9) (September 1967).
21. Johnson, R. A., Kast, F. E., and Rosenzweig, J. E., *The Theory and Management of Systems.* New York: McGraw-Hill, 1967.
22. Desmonde, W. H., *Computers and Their Uses.* Englewood Cliffs, New Jersey: Prentice-Hall, 1964.
23. Thesen, Arne, *Computer Methods in Operations Research.* New York: Academic Press, 1978.
24. Ackoff, R. L., Some Ideas on Education in the Management Sciences, *Management Science, Appl. Ser.* 17 (2) (October 1970).
25. Churchman, C. W., Perspectives of the Systems Approach, *Interfaces* 4 (4) (August 1974).

RECOMMENDATIONS FOR FURTHER READING

Ahlers, David M., An Investment Decision Making System. *Interfaces* 5 (2), 72–90 (February 1975).
Back, Harry B., A Comparison of Operations Research and Management Science, *Interfaces* 4 (2), 42–52 (February 1974).
Berlin, Victor N., Administrative Experimentation: A Methodology for More Rigorous "Muddling Through" *Management Science* 24 (8), 789–799 (April 1978).
Burt, John M., Planning and Dynamic Control of Projects Under Uncertainty, *Management Science* 24 (3), 249–258 (November 1977).
Cyert, R. M., DeGroot, M. H., and Holt, C. A., Sequential Investment Decisions with Bayesian Learning, *Management Science* 24 (7), 712–718 (March 1978).
DeBrabander, Bert, and Edstrom, A., Successful Information System Development Projects, *Management Science* 24 (2), 191–199 (October 1977).

Derman, C., Lieberman, G., and Ross, S. M., A Renewal Decision Problem, *Management Science* 24 (5), 554–561 (January 1978).

Eldin, H. K., Education in Systems Engineering, *Engineering Education* 60 (8) (April 1970).

Eliashberg, Jehoshus, and Winkler, Robert L., The Role of Attitude Toward Risk in Strictly Competitive Decision-Making Situations, *Management Science* 24 (12), 1231–1241 (August 1978).

Fuller, Jack A., and Atherton, Roger M., Fitting in the Management Science Specialist, *Business Horizons* 22 (2), 14–17 (April 1979).

Greco, Richard J., MIS Planning—An Approach, *Data Management* 9 (10), 17 (October 1971).

Gruber, William H., and Niles, John S., The Science–Technology–Utilization Relationship in Management, *Management Science* 21 (8), 956–963 (April 1975).

Gupta, Jatinder N. D., Management Science Implementation—Experiences of a Practicing O.R. Manager, *Interfaces* (*TIMS*) 7 (3), 84–90 (May 1977).

Helmer, E. D., The Prospects of a United Theory of Organizations, *Management Science* 4 (2) (January 1958).

Hoffman, Gerald M., The Contribution of Management Science to Management Information, *Interfaces* (*TIMS*) (1), 34–39 (November 1978).

Konczal, Edward F., Documenting Large-Scale Telecommunications Computer Analyses, *Journal of Systems Management* 29 (6), 14–17 (June 1978).

Lindley, D. V., *Making Decisions*. New York: Interscience, 1971.

Lorsch, Jay W., Making Behavioral Science More Useful, *Harvard Business Review* 57 (2), 171–180 (March/April 1979).

Milliken, Russell B., Tax Professionals at Harvard Business School—Rewarding Experience for Mead's Director, *Tax Executive* 28 (3), 243–252 (April 1976).

Montanari, John R., Managerial Discretion: An Expanded Model of Organizational Choice, *Academy of Management Review* 3 (2), 231–241 (April 1978).

Olson, Philip D., Notes—Decision-Making Type 1 and Type 2 Error Analysis, *California Management Review* 20 (1), 81–83 (Fall 1977).

Oppenheimer, Kenneth R., A Proxy Approach to Multi-Attribute Decision Making, *Management Science* 24 (6), 675–689 (February 1978).

Pike, Dan, Management Theory: Its Application to the Job, *Supervisory Management* 23 (12), 26–30 (December 1978).

Render, Barry, and Stair, Ralph, Management Science and the Small Business, *Journal of Systems Management* 28 (3), 20–22 (March 1977).

Riggs, James L., and Irone, Michael S., *Introduction to Operations Research and Management Science*. New York: McGraw-Hill, 1975.

Schneyman, Arthur H., Management Information Systems for Management Sciences, *Interfaces* (*TIMS*) 6 (3), 52–59 (May 1976).

Schwarz, Leroy B., and Johnson, Robert E., An Appraisal of the Empirical Performance of the Linear Decision Rule for Aggregate Planning, *Management Science* 24 (8), 844–849 (April 1978).

Sprague, Linda G., and Sprague, Christopher R., Management Science, Part 1, *Interfaces* (*TIMS*) 7 (1), 57–62 (November 1976).

Thompson, Gerald E., *Management Science, An Introduction to Modern Quantitative Analysis and Decision Making*. New York: McGraw-Hill, 1976.

Vazsonyi, Andrew, Information Systems in Management Science Data—Base Management Systems, *Interfaces* 5 (3), 47–52 (May 1975).

Vazsonyi, Andrew, Information Systems in Management Science–Decision Support Systems: The New Technology of Decision Making, *Interfaces* (*TIMS*) 9 (1), 72–77 (November 1978).

Wagner, Harvey M., *Principles of Management Science*. Englewood Cliffs, New Jersey: Prentice-Hall, 2nd ed., 1975.

Chapter 2
Management Science and Modeling Techniques

INTRODUCTION

Managerial problems manifest themselves in many ways. Some have been around for a long time and are dealt with in a makeshift manner rather than recognized or isolated as problems. Other problems more directly affect the objectives of the organization, and are recognized as such in a more immediate fashion. However, the manner in which a problem comes to be recognized has very little to do with the speed or the overall success of the solution. Reliability of the solution is dependent upon how accurately the technique of solution applies to the type of the problem. Managerial problems sometimes indicate strongly which technique would be best, but often these problems are difficult to classify as to solution technique. In this chapter we present a general approach as to the matching of a problem with the technique best qualified to provide an accurate solution to that problem.

In order to find the best solution to a problem, we should follow certain basic steps:

Identify the problem by content

Identify the problem by form

Define variables of the problem

Develop the model

Choose the solution category

Choose the right technique

I. PROBLEM CONTENT AND FORM

A. Identify the Problem by Content

The transition between recognition of a problem and actual identification of the problem itself is dependent upon the scope of the problem and the accuracy of its information feedback. If the scope of the problem is narrow, then the job of identifying all aspects of the problem is simplified. As the breadth of the scope increases, the job of identification becomes more complex, in which case there are two avenues of approach. One is that the problem be broken up into several subproblems that are narrower in scope. The other is that the problem may be considered in terms of strategic elements only. Some problems are almost synonymous with the major departments of a large organization and are classified as problems in research and development, production, marketing, finance and accounting, industrial relations, or logistic applications. It is obvious that not all of management's problems will fall neatly into the above categories; many will tend to overlap and some may even affect the whole structure of the organization. Once a problem has been identified by content, it must be identified by form.

B. Identify the Problem by Form

It is generally accepted that most management content problems exhibit features that allow them to be classified into one or more of these forms: inventory, allocation, queuing or waiting line, routing, sequencing, replacement, competition, and search.

Inventory: Problems that are of this form have the common element of maintenance of some physical item that has an economic value to meet future demand. Special techniques to minimize the cost associated with this problem have been developed. However, since future demand generally has some degree of uncertainty associated with it, problems of this type will probably best be solved by a stochastic technique.

Allocation: These problems have in common the desire to maximize the profit or minimize the cost as an objective function with constraints imposed upon it. These problems generally lend themselves to an optimization model.

Queuing or waiting line: Problems of this form generally have in common a congestion due to the irregularities in the system. These irregularities may be caused by an irregular arrival of units to be serviced, an unexpected length of time required to service individual units, or some combination of the two. This is another form of problem that must consider future events, thus indicating a stochastic technique of solution. In fact there is an extensive application of probability theory in solving waiting line problems.

Sequencing: This type of problem occurs when several jobs require a common machine. The desire to use the machine efficiently and maximize return from the individual job leads to a certain sequence in which those jobs are performed. Optimization techniques are most helpful.

Routing: These problems have to do with making a prescribed number of stops in a minimum amount of distance. Dynamic programming and branch and bound techniques are two powerful techniques that can be used for problems of this form.

Replacement: Problems of replacement of machines, materials, or employees are generally solved by an economic analysis; the replacement occurs if the cost of operation, maintenance, and obsolescence is greater than the average total cost. In economic analysis of this type there are some variables that are deterministic and some that are stochastic. The distinction between the two must be made in advance.

Competition: In problems of competition, two or more persons or groups are competing for the same objective, for which there is a conflict. A decision maker must be aware of the choices available to him and to his competitor, as well as the payoffs associated with his decision. These problems are usually dealt with by the theory of games.

Search: The problem of searching exists when there is a need to determine the value or property of a particular datum from among a large list of data. For example, such a search is conducted when you look for a subject in the subject index of a book. Search techniques are capable of solving complex problems. However, there are fewer techniques available for solving search problems than there are for other types. Search techniques, in general, do not guarantee an optimum solution to the problem.

C. Define Variables of the Problem

In order to develop a model for a problem, we need to identify and define its elements or variables. These variables should give accurate representation of the decision to be made and provide the management with an appropriate course of action to consider in order to attain its goals.

Once the variables of the problem are defined, then the relationships among the variables must be expressed so that mathematical equations and functions can be developed.

Variables of a problem or a system are usually categorized into one of the two groups: controllable and uncontrollable. *Controllable* variables are those that could be brought under control by the decision maker and can be changed by changing policies or strategies. These variables are often referred to as decision variables. Decision variables could be dependent on or independent of some events that are related to the problem. For example, in a manufacturing firm, the price of raw materials and the inventory carrying cost are considered uncontrollable and controllable variables, respectively. If the problem is to minimize the total cost of operation, these two variables will be part of the decision variables of the problem. The price of raw materials is an independent variable in this case, whereas the inventory carrying cost is a dependent variable, since the cost will be dependent upon the number of units in the stock.

Uncontrollable variables are those over which the decision maker has very limited or no control. Again, these variables could be dependent on or independent of events surrounding the problem. However, these variables affect the solution of the problem and should be considered and incorporated into the model when possible. Demand, inflation rate, interest rates, and government regulations are a few examples of *uncontrollable* variables.

II. MODEL DEVELOPMENT

Models, like the problem itself, can be simple or complex. In management science we use various types of model, varying in degree of complexity. Models, in general, could be classified as follows:

Analog or schematic models: These represent a pictorial or graphical relationship among the variables of the system. The behavior of these models is the same as the real system although they may not physically resemble it. Blueprints of a building or an engine, city maps, and organization charts are examples.

Iconic models: These are physical representations of the real system scaled up or down. For example, a toy gun or a model airplane is an iconic model of a real gun or an airplane.

Descriptive models: These describe the system or the problem and represent the activities or conditions of the system or the problem. They are used as a framework that would provide information to the decision maker. Descriptive models are usually in verbal form, such as asking directions, or in written form, such as blueprints, maps, and charts.

Mathematical or symbolic models: Sets of mathematical symbols, equations, inequalities, and expressions used to show the functional relationships among decision variables in order to describe the behavior of the real system effectively.

After developing the model, either the solution technique will be apparent, or the general classification of the technique will be realized. It is necessary to note here that at any point in the process of choosing and/or applying the solution technique, it is imperative to review and perhaps refine the model. This evaluation involves an analysis of the parameters used with specific attention to the following:

1. Addition of previously ignored ones
2. Elimination of unnecessary ones
3. Extensive questioning of the correctness of existing parameters

III. SOLUTION CATEGORY AND TECHNIQUE

A. Choose the Solution Category

As mentioned in Section II, after the development of the model, the technique may be apparent; if it is, the steps described in Sections III and IV may be skipped. However, the management scientist must be especially careful to choose the technique to fit the model and not the reverse. Sometimes, the particular technique is not obvious, but the general solution category is. There are two general classifications: deterministic or stochastic.

B. Choose the Right Technique

If a problem lends itself to solution by a deterministic technique, then a further classification is necessary between two complementary subsets: techniques with unique solutions and those with optimal solutions.

If it is ascertained that a stochastic model is necessary, then it should become clear that there exists a situation in which some of the unknowns can be said to occur with a certain probability. The parameters of stochastic models thus acquire the added dimension of probabilistic influence.

IV. TECHNIQUES OF MANAGEMENT SCIENCE

There are many management science tools and techniques available for solving either deterministic or stochastic problems. Broad classifications of these tools and techniques are outlined here; more detailed applications and discussions are presented throughout the book.

A. Optimization Techniques

Optimization techniques are utilized in situations where there are many feasible solutions and the objective is to find the optimal one.

The most basic analytical tool of optimization is *differential calculus*. For an unconstrained function with a single variable, relative maxima or minima may be determined by applying first- and second-derivative tests. This procedure may be generalized in certain cases to include constrained functions of multiple variables. Optimal values are found in terms of artificial parameters, and proper values satisfying the constraints are then chosen. However, the usefulness of this procedure decreases rapidly as the size and the complexity of the problem increases.

Another optimization technique is the *gradient method*. The gradient of a function is the vector of partial derivatives of that function. When the gradient is evaluated at a particular point, it indicates the direction of maximum increase of the function. Thus, through a series of iterations of gradient search procedure, it is possible to determine a local maximum or, in some special cases, a global maximum. If the objective is to minimize, then we should move in the opposite direction of the gradient each time we iterate.

Another class of optimization technique is *mathematical programming*. It is inherently a powerful tool for analyzing optimization problems since it can cut across the enormous combinational range of alternatives by providing a systematic search for the optimal solution. The general form of a mathematical programming problem is that there is a desire to maximize or minimize an objective function subject to constraints on the variables. The constraints refer to the range of values for each variable. The intersection of all the individual spaces represented by the constraints is called the *solution space*. Once this is found, the objective function is evaluated for each restricted value of the variables and the maximum or minimum value is decided.

B. Utilization of Computers

The electronic computer is probably the most powerful tool for progress that we have ever had at our disposal. Many decision-making problems are rather complex and require complicated procedures for solution. These problems are usually time consuming and sometimes impossible to solve without the use of a computer.

In the field of formulating goals or establishing policies the role of computers is small. This is a field in which human judgment is indispensable. However, in order to develop policies or goals, many details are required and many routine steps and calculations must be carried out. Here is where the utilization of computers is beneficial.

Many executives are presently burdened with making many routine decisions. If relieved of this responsibility, the executive's time, imagination, and intuition are free for the more valuable work that the computer cannot do, such as setting objectives, policy, and handling such problems as labor negotiations. Whole new branches of computerized mathematical techniques have been developed for scheduling production, controlling inventory, figuring out warehousing and distribution networks, and determining the optimal location of new facilities, to mention a few.

C. Heuristic Models

In considering the concept of heuristic problem solving, algorithms are not available, so solutions are not ensured. Random search techniques are sometimes impractical. Therefore, a selected search technique is employed by recognizing patterns of improvements in solutions. As the problem is solved, the method becomes more and more efficient. Included in the heuristic approach is the *pattern search method*. Pattern search is composed of two basic operations: the exploratory search and the pattern move. The exploratory search is done in order to determine a successful direction. Whenever the exploratory search has been successful, a pattern move is made. Then the pattern move grows and changes direction with the successes in the exploratory search. Acceptance of the heuristic or similar technique is a compromise to obtain a solution within reasonable cost and elapsed time.

D. Simulation

Simulation is another powerful tool available to the decision maker. Proper simulation requires a close approximation to reality. If such an approximation does not exist in the simulation model, then the model should appropriately be called statistical or probabilistic. The mathematical quantities should be related in such a way that they bear direct resemblance to the real-world situation. Simulation is used to test ideas and proposed changes in advance or compare alternative courses of action.

For purposes of long-range planning, the overall condition of the industry must be taken into account. The *industrial dynamics* simulation model is of a sort applied to the total perspective of the business organization. This type of model has been used to simulate the growth behavior of business organizations, the implications of a change in marketing strategy or in products and product mix, the fluctuations of a price and supply in commodity markets, and other situations. Harmony must be achieved between the parameters and the internal variables before any type of simulation technique will be reasonably accurate.

One of the most prevalent forms of simulation is the *business game*. The computer serves as a central data processor and, given the periodic decisions of the individuals or teams playing the game, computes the interaction of these decisions within firms and among firms and reports the outcome to those playing the game. In this way, competitive business situations calling for the development of policies and decision making are simulated. Business games exist for various levels of complexity and sophistication. However, it must be remembered that simulation fundamentally rests on a foundation of probability and statistics. Without faith in probability, none of the simulation models is of any use.

V. THE SYSTEMS APPROACH

With the growth complexities and diversity of operations in today's advanced technology environment, it is extremely important for managers to be able to think on an overall system basis, integrating numerous complex operations.

A *systems approach* calls for the systematic organization of very large numbers of technologies, industries, human and material resources, and other subsystems into an integrated whole toward the accomplishment of certain objectives. This could be thought of as an interspersion of systems methodology with the utilization of proper tools and techniques.

From the data on day-to-day operations of the business, the manager and the operator extract information as to trends occurring in the business. Only by analyzing this data systematically, rapidly, and comprehensively can they gain control over the operations and achieve a sound basis for subsequent planning and decision making. The computer makes it possible to know what is going on and why, the alternative plans and their consequences, and the courses of action that could or should be taken. It increases a manager's span of effective control, with greater summarized detail, and permits rapid action. All of this is achieved only if the application is planned, programmed, and within the capability of the computer.

In examining a typical problem-solving method, it is not enough to list various subprocesses; they must be interrelated and the structure of each one understood. A person may start with problem definition, the logical conclusion of which is the setting of definite objectives. Objectives serve to indicate the types of analysis required of the alternative systems synthesized and provide the criteria for selecting the optimal system. *Systems synthesis* includes compiling alternative systems that can satisfy the objectives. *Systems analysis* entails deducing the consequences of an entire list of hypothetical systems. Selecting the best system involves evaluating the analysis and comparing that evaluation with the objectives. Planning for execution of the chosen alternative includes designing a time-associated schedule and setting up an appropriate organization, reporting systems, control measures, information flow systems, and so on.

Information originating in the operations at lower levels of the organization is communicated to higher levels which, in turn, produce command decisions that are transmitted to the lower levels. The flow of information and decisions necessary to the operation of any organization, whether or not it incorporates computers, follows along the same lines. When computers are introduced, decisions at all levels can be greatly expedited, routine decisions automated, operations optimized, and management information transmitted in real time, while it is still current and useful, rather than after it has become history. The last statement is probably the most important aspect of this type of management system. Current and factual information is extremely important for fast correct decision making. Indeed, this is an important aspect of the scientific management era.

As an example of this management science technique, let us examine a petroleum company that has set up a system of computer control at the refinery level. This system provides for refinery planning and decisions, such as the scheduling of processing of various crudes to meet changing market demands. A setting of daily targets of operation for each processing unit has been facilitated by this system. It is adapted to tie in with a master computer at the corporate level, where it will aid such corporate decision making as establishing market requirements, selecting crudes, and determining the quantities and processing schedules for the individual refineries.

This concept of overall control, the total systems approach, is very important today. However, its use requires the design of total management operating and information systems. This tremendous task will fall to those people who are skilled in the science of management.

REVIEW QUESTIONS

1. Problem solving as a process involves several basic steps. Explain these steps.

2. Contrast problems "by content" and "by form."

3. Give three examples that show how queuing theory can be used to aid corporate managers.

4. Define "controllable" and "uncontrollable" variables. Explain how each can affect the solutions of a model.

5. What is the function of a model in a management science problem? In what ways can a model be useful?

6. Describe the various types of model that are available to management.

7. What are the differences between optimization models and heuristic models?

8. List and explain the types of decision that should be computerized.

9. Why does a management scientist need computers?

10. List four managerial problems that could be solved by management science techniques. List some that could not.

11. Why is the total systems approach important to management?

RECOMMENDATIONS FOR FURTHER READING

Beer, S., *Management Science: The Business Use of Operations Research*. Garden City, New York: Doubleday, 1968.

Bonini, Charles P., Computers, Modeling, and Management Education, *California Management Review* 21 (2), 47–55 (Winter 1978).

Brinkloe, W. D., *Managerial Operations Research*. New York: McGraw-Hill, 1969.

Brown, Gerald G., and Rutemiller, Herbert C., Means and Variances of Stochastic Vector Products with Applications to Random Linear Models, *Management Science* 24 (2), 210–216 (October 1977).

Butler, David A., A Hazardous-Inspection Model, *Management Science* 25 (1), 79–89 (January 1979).

Chaiken, Jan M., Transfer of Emergency Service Deployment Models to Operating Agencies, *Management Science* 24 (7), 719–731 (March 1978).

Dokmeci, Vedia F., A Quantitative Model to Plan Regional Health Facility Systems, *Management Science* 24 (4), 411–419 (December 1977).

Erlenkotter, Donald, Facility Location with Price-Sensitive Demands: Private, Public, and Quasi-Public, *Management Science* 24 (4), 378–386 (December 1977).

Esogbue, Augustine O., and Marks, Barry R., Dynamic Programming Models of the Nonserial Critical Path-Cost Problem, *Management Science* 24 (2), 200–209 (October 1977).

Gardner, Bert, Electrify Your Thinking, *Accountancy* 85 (976), 86–89 (December 1974).

Gupta, Shiv K., A Language for Policy-Level Modelling, *Journal of the Operational Research Society* 30 (4), 297–308 (April 1979).

Higgins, J. C., and Finn, R., Managerial Attitudes Towards Computer Models for Planning and Control, *Long-Range Planning* 9 (6), 107–112 (December 1976).

Lilien, Gary L., and Rao, Ambar G., A Model for Allocating Retail Outlet Building Resources Across Market Areas, *Operations Research* 24 (1), 1–14 (January/February 1976).

McMillan, C., and Gonzalez, R. F., *Systems Analysis: A Computer Approach to Decision Models*. Homewood, Illinois: Richard D. Irwin, 1965.

Mjosund, A., The Synergy of Operations Research and Computers, *Operations Research* 20 (5), 1057–1064 (1972).

Morris, W. T., On the Art of Modeling, *Management Science* 13 (2), 707–717 (August 1967).

Paul, Robert J., The Retail Store as a Waiting Line Model, *Journal of Retailing* 48 (2), 3–15 (Summer 1972).

Ross, G. Terry, and Soland, Richard M., Modeling Facility Location Problems as Generalized Assignment Problems, *Management Science* 24 (3) (November 1977).

Shycon, Harvey N., All Around the model—Perspectives on MS Applications, *Interfaces* 5 (1), 41–43 (November 1974).

Steudel, Harold J., Pandit, S. M., and Wu, S. M., A Multiple Time Series Approach to Modeling the Manufacturing Job-Shop as a Network of Queues, *Management Science* 24 (4), 456–463 (December 1977).

Wade, James C., and Heady, Earl C., A Spatial Equilibrium Model for Evaluating Alternative Policies for Controlling Sediment from Agriculture, *Management Science* 24 (6), 633–644 (February 1978).

Welam, Ulf Peter, An HMMS Type Interactive Model for Aggregate Planning, *Management Science* 24 (5), 564–575 (January 1978).

Wilde, D. J., and Beightler, C. S., *Foundations of Optimization*. Englewood Cliffs, New Jersey: Prentice-Hall, 1967.

Chapter 3
Modeling for Decision Making

INTRODUCTION

A decision is the selection of one course of action from two or more possible alternatives. The process of deciding is methodical; it includes analysis, synthesis, and selection of one of several possible solutions on the basis of a criterion. This chapter is concerned with the process of decision making under circumstances ranging from certainty to complete ignorance of the possible consequences of a decision.

I. FACTORS AFFECTING DECISIONS

The decision maker should consider the following aspects of the decision-making process in an organized business analysis.

Explicitness of estimates of the future: In order to have reliable estimates of future events, all collected data should be understood clearly and tested for degree of validity.

Reasonable alternatives: The decision maker should search for all reasonable alternatives and consider them equally in an organized, thorough way.

Weaknesses and Gaps: Very often in a complex business situation, a gap in the data or logic can slow the analysis without being noticed. There are specific ways to discover this, as well as methods to determine the sensitivity of all parts of a situation to a weakness or change in any one part.

Modeling and testing alternatives: It is possible not only to model alternatives in a situation but, also equally important, to test their implications against various combinations of estimates of the future.

Time shortening: The combination of analytical methods and computer speed can significantly shorten the time span between problem inception, analysis, decision, and payoff.

Communication: One of the least emphasized and most important needs in complex situations is the ability to structure and communicate all the pertinent information, assumptions, alternatives, and questions that apply.

Iteration: Finally, the best possible solutions to meet present goals can be determined by repeated manipulation of all of the information.

Optimal consideration of these factors involves using human judgment and quantifying it to a far greater degree than in conventional business analysis. It also requires conscious application of a logical approach and appropriate selection of the available tools.

The tools available for use in complex problems are really extensions of some of the processes a decision maker would use for simple problems; they merely organize perspective so decisions can be made. The underlying ideas are more or less universal and founded on good common sense.

The choice of mathematical models, tools, and techniques to be used in decision making depends upon how the decision is classified and what information is known about the problem. *Decision theory* is a technique that can be used to analyze a problem and come up with a set of alternative solutions. In each decision situation, there are alternative courses of action that may be taken. Associated with each course of action or alternative, there is one or more possible resultant states, called the *potential states of nature* for the problem. Depending upon what is known about the potential states of nature, the decision-making process may be classified as being under certainty, risk, uncertainty, or conflict.

II. DECISION UNDER CERTAINTY

To say that a decision is made under the condition of certainty implies that there exists only one feasible state of nature for each possible course of action. Once a decision is made, the future is known, stable, and is unalterable as far as the decision maker is concerned. Decision would then consist of computing the value of the outcome for each alternative and selecting that associated with the most desirable outcome. In other words, decisions involving certainty are characterized by the complete knowledge of a system, which eventually results in only one possible sensible outcome: that having the highest return or the lowest investment.

When decisions involve selecting a value for a single variable and the return on the decision is the function of the decision variable, maximization or minimization techniques from calculus may be used in order to find the best alternative. Break-even analysis, depreciation models, and certain types of inventory model are a few examples of how decisions under certainty may be treated.

III. DECISION UNDER RISK

Decisions under risk implies that for each course of action there is more than one possible resultant states of nature each of which can be assigned a likelihood (probability) of occurrence. For example, when a new product is introduced, product demand is uncertain but can be estimated with some probability distribution. Many tools are available for use in such situations, depending upon the particulars available.

Expected value analysis is commonly used in this type of problem. Originating from statistical analysis, expected value is simply a weighted average of the measure of effectiveness of the possible outcomes of a decision. This technique is discussed more thoroughly in Chapter 8, Section II.G.

Other techniques such as Bayes' theorem, decision tree, and Markov chains can also be used in decisions under risk.

A. Probability Theory and Bayes' Theorem

Process selection, planning, and the actual operation involve many instances where the manager must utilize elements of probability theory to formulate decisions. The inputs to probability evaluations usually take the form of observations made on the present environment or a similar one. These observations provide the basis for further inferences about the situation. We shall consider two types of probability in this context: objective and subjective.

The probability that an event will occur is the ratio of the chances favoring an event to the total number of chances both for and against the event. The first task in considering the future through the application of probabilities is to obtain estimates of the likelihood of the events in which we are interested. The source of these estimates may be of either an objective or subjective nature.

Objective evidence of probability is usually in the form of historical documentation or common experience. For example, it is readily accepted that a coin has two sides and a die has six. A logical person would not question the assignment of a probability of $\frac{1}{2}$ to the occurrence of a head on the flip of a coin or a probability of $\frac{1}{6}$ to a "2" from the roll of a die. This type of information is known as *prior knowledge* and the probability associated with this knowledge is called *prior probability*. Subjective probability estimates are derived from opinions based on general experience and knowledge that pertains to the particular situation under consideration. A subjective probability represents the extent to which an individual thinks a given event is likely to occur. Subjective probabilities have the same mathematical properties as objective probabilities but reflect an individual's viewpoint. Individuals are free to choose any subjective probability they like prior to the first occurrence of an event; after it, the subjective probability changes, as a result of gained experience, in a manner governed by Bayes' theorem. Therefore, two individuals may start out with very different subjective probability estimates of the same event, but after a number of occurrences their subjective probabilities will tend to converge. The use of subjective probability is of extreme importance to the decision maker, since it is the variety most often available.

Before venturing into the rules of probability, we must define some basic terminologies. The selection of a single card from a deck of cards is an *event*. All the different possible events (52 in a deck of cards) taken together compose a *population*. The population can be divided into *sets*, such as the set of 13 hearts or diamonds.

Events can be statistically dependent, independent, or mutually exclusive. *Independent events* are those unaffected by the occurrence of any other event. The probabilities that two or more independent events will occur together or in succession is the product of the individual probabilities of all the events involved.

If two or more events cannot happen at the same time, they are called *mutually exclusive*. The probabilities of mutually exclusive events can be added. The probabilities of drawing a queen from our deck of cards is the sum of the probabilities of all four queens in the deck.

An event is said to be *statistically dependent* when its outcome is affected by the occurrence of another event. We can best illustrate dependent conditions by an example. Consider two boxes, labeled X and Y, containing black and white marbles. Box X contains three white and three black marbles, box Y holds one white and five black marbles. The probability of drawing a white marble from *either* box is $(\frac{3}{6} \times \frac{1}{2}) + (\frac{1}{6} \times \frac{1}{2}) = \frac{1}{3}$, which is the sum of the white-marble probabilities for boxes X and Y taken individually. This is also called the marginal probability of drawing a white marble.

To introduce statistical dependence, let us examine the probability that a white marble will be drawn from box Y, which can be expressed symbolically as $P(W|Y)$, where the vertical line is read "given." Obviously, there would be one chance in six that a white marble would be drawn from box Y. However, given the choice ("occurrence") of the conditional probability $P(W|Y)$ that a white marble will be drawn, box Y is equal to the probability $P(WY)$ that a white marble will occur in box Y divided by the probability $P(Y)$ that the draw will be made from box Y:

$$P(W|Y) = P(WY)/P(Y)$$

$P(WY)$, where WY denotes the set of points simultaneously in W and Y, represents the joint probability of two (in this case, independent) events W and Y. In our example the joint probability that a marble will be white and in box Y is

$$P(WY) = P(W) \times P(Y) = \tfrac{4}{12} \times \tfrac{1}{2} = \tfrac{1}{6}$$

Thus

$$P(W|Y) = \tfrac{1}{6}/\tfrac{1}{2} = \tfrac{1}{3}$$

Let $P(E)$ be the probability of a simple event that is unrelated to the occurrence of any other event, although such unrelated events are rare. For example, if A, B, and C denote possible choices and a_1, a_2, a_3, . . . , a_n denote a set of outcomes related to choice A, then the probability of a given outcome a_1 depends on the prior occurrence of A. We may write $P(a_1|A)$ to represent the full choice situations, that is, the probability of a_1 given that A is selected from a group of possible choices (A, B, C). The *marginal* probability $P(a_1)$, the measure of the occurrence of a_1 without reference to A, is appropriate only if the probability of a_1 is independent of the choice of A.

As defined earlier, the prior probability of an event may be modified by experience gained through experimentation. The controlling element in this area is usually Bayes' theorem. Bayes' basic equation can be derived as follows. Start with the general form of the equation for conditional probabilities,

$$P(i|j) = P(ij)/P(j)$$

which can be rewritten as

$$P(ij) = P(i|j) \times P(j)$$

since ij and ji represent the same set, and

$$P(ji) = P(j|i) \times P(i)$$

we can substitute to yield

$$P(i|j) \times P(j) = P(j|i) \times P(i)$$

which can be converted to

$$P(i|j) = \frac{P(j|i) \times P(i)}{P(j)}$$

which is Bayes' theorem. The marginal probability of j is

$$P(j) = P(j|i_1) \times P(i_1) + P(j|i_2) \times P(i_2)$$

$$+ \cdots + P(j|i_n) \times P(i_n)$$

$$= \sum_{x=1}^{n} [(j|i_x) \times P(i_x)]$$

This allows event i to be reevaluated when new information concerning the outcome of j becomes available.

Probability theory is one of the most frequently used tools available to the decision maker. The effective use of this tool provides the decision maker with an estimate of future events, a base from which the decision maker may plan future actions. However, it should be remembered that probability figures can be no more accurate than the data upon which they are built. It is therefore most important to obtain the best initial data possible.

B. Decision Tree Analysis

Decision tree methodology depends upon recognition of future alternatives, possible outcomes, and decisions that can result from an initial decision. Problems or decisions that involve a series of time-phased stages can be analyzed by decision tree techniques in order to find the best alternative course of action.

To consider random outcomes, probabilities of occurrence must be introduced; should further information on probabilities become available, Bayesian theory can be utilized to update the decision tree. After the tree is set up with alternatives, times, resources, probabilities, and so on, expected outcomes can be calculated. The desirability of the various outcomes is then examined by the criterion of expected utility.

In order to construct a decision tree, we need first to determine all the available alternatives and then identify the likelihood of occurrences for each. Figure 1 shows the general structure of a decision tree with the stages of decisions and the associated probabilities. Each branch designates a decision alternative which joins either two nodes or a node to an outcome. Decision points are indicated by squares and chance points by circles. A chance point designates the existence of uncontrollable events. Every time a chance point is added to the tree, the related states of nature with their corresponding probabilities are also added. The process of adding decision and chance points continues until expected outcomes are reached. It should be noted that although a time scale is not represented by the length of branches in the tree, the tree does depict chronological

52

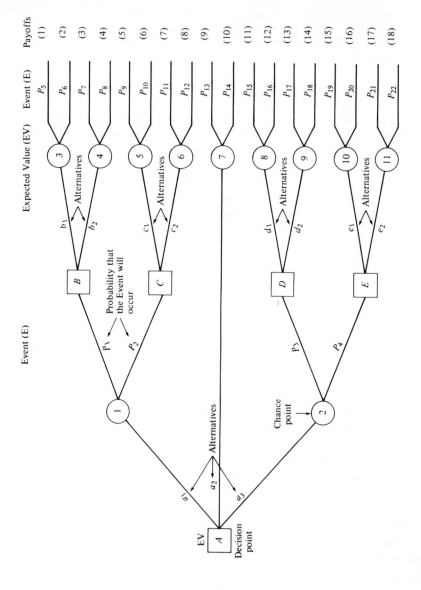

FIGURE 1. General structure of a decision tree.

order. Starting from the right, expected values of all chance points are computed; the alternative with highest expected value is selected.

As an example, suppose that a manufacturing firm is considering whether or not it should expand its facilities for the next two years. A survey of the market shows that there is a 60% chance of high demand for the company's product next year. The probability of high demand for the second year is estimated to be 70%. Thus, there exists a chance of 40% for the first year and 30% for the second year that there will be a low demand (assuming all demand values to be either high or low). The company has two alternatives: either expand the existing facilities at a cost of $30,000 or add an additional shift at a cost of $10,000.

At the end of the first year the company is faced with the following alternatives:

1. If expansion had been the choice and a high demand had been experienced, then management could either further expand the operations or establish an additional shift. The predicted revenues are $120,000 and $80,000, respectively, if expansion is the choice. If an additional shift is established, the revenues will be $90,000 and $75,000 for high and low demand, respectively.

2. If expansion had been the choice but resulted in low demand, then the management would consider the use of existing capacity, yielding $75,000 revenue in the case of high demand and $60,000 for low demand.

3. If an additional shift had been established initially and demand had been high, then the decision maker would have two alternatives: expand the present facility or institute overtime for every shift, which would result in a total cost of $20,000. The anticipated revenues for high and low demands are $75,000 and $60,000, respectively, for the overtime alternative but $90,000 and $75,000, respectively, in the case of expansion.

4. If an additional shift had been established initially and a low demand experienced, then the management would continue with the additional shift. The estimated revenues for high and low demands are $65,000 and $55,000, respectively.

The objective is to find the best alternative for the company at the present time and at the end of the first year, assuming that all cost and revenue figures are in present value.

In order to solve the problem, we need first to construct a decision tree. This tree is shown in Figure 2. The next step, starting from the right, is to compute the expected revenue (ER) at all chance points:

Node 3: $ER = 120,000 \times .7 + 80,000 \times .3 = \$108,000$

Node 4: $ER = 90,000 \times .7 + 75,000 \times .3 = \$ 85,500$

Node 5: ER = 75,000 × .7 + 60,000 × .3 = $ 70,500
Node 6: ER = 90,000 × .7 + 75,000 × .3 = $ 85,500
Node 7: ER = 75,000 × .7 + 60,000 × .3 = $ 70,500
Node 8: ER = 65,000 × .7 + 55,000 × .3 = $ 62,000

Moving leftward, we arrive at decision points *B, C, D,* and *E.* At this stage we need to compute the expected net revenue (ENR) for every decision point by utilizing the following equation:

Expected net revenue = Expected revenue − Cost

Therefore, the expected net revenue for each decision point is as follows:

Node B: In case of expansion the expected net revenue is $108,000 − $30,000 = $78,000 as compared with the alternative of an additional shift which yields $85,500 − $10,000 = $75,500. Since the first alternative is more profitable, it is selected and is assigned as the value of decision point *B.*

Node C: At this point, since there is only one alternative, the expected revenue of $70,500 is assigned to the value of point *C.*

Node D: There are two alternatives at this decision point: expansion of facilities with an expected net revenue of $55,500 ($85,500 − $30,000) or overtime for every shift, which would result in a net expected revenue of $50,500 ($70,500 − $20,000). Since the first alternative would yield a higher expected net revenue, it is selected; thus $55,500 is assigned as the value at *D.*

Node E: Again there is only one alternative at this decision point, so an expected net revenue of $62,000 is considered for decision point *E.*

Now only the left side of the tree remains, as shown in Figure 3.

Following the same procedure as before, expected revenues for chance points 1 and 2 can be computed as follows:

Node 1: ER = 78,000 × .6 + 70,500 × .4 = $75,000
Node 2: ER = 55,500 × .6 + 62,000 × .4 = $58,100

In order to find out which alternative should be accepted at decision point *A* we need to compare the expected net revenue of each alternative and choose that with the highest yield. In the case of expansion, the expected net revenue is $45,000 ($75,000 − $30,000); for the case of an additional shift, it is $48,100 ($58,100 − $10,000). This indicates that at decision point *A* the additional shift is a better alternative than expansion.

Therefore, for the first year it is better to add an additional shift. For the second year, if there is a high demand in the first year, it is advisable

55

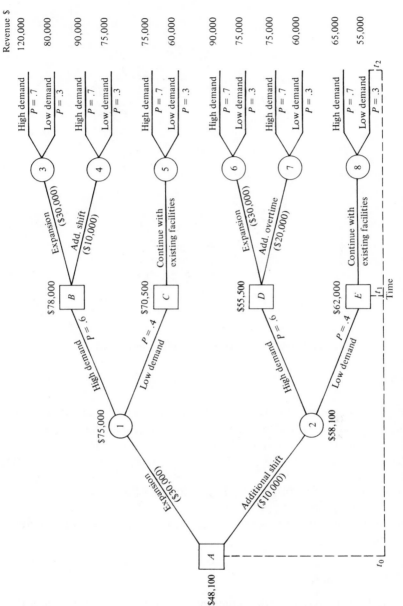

FIGURE 2. Decision tree for the manufacturing company.

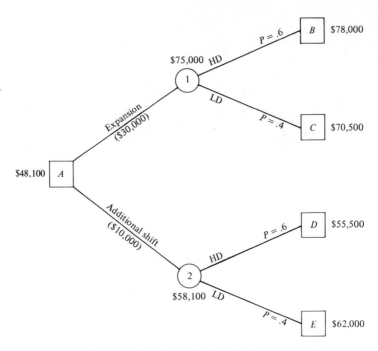

FIGURE 3. Decision tree for the first year.

to expand the operations. However, in the case of low demand in the first year, the management should continue with their existing facilities.

The decision tree technique provides management with a clear graphical presentation of the impact of alternative courses of action by utilizing expected value as a tool of analysis. However, it should be recognized that when a decision involves a large number of alternatives and states of nature the graphical presentation of a decision tree becomes rather time consuming and laborious. In such cases, the use of a computer is required.

C. Markov Chains

The Markov chain is another technique for analyzing decisions under risk. A Markov chain may describe a system or process that is stochastic in nature and for which the conditional probability of any future event, given any past event and the present state, is independent of the past event and depends only upon the present state of the process. The model for such a system is generally represented as a matrix of the probabilities of future events of states, known as a *one step* or *transition matrix*. The matrix shown in Figure 4 represents the general form of the transition matrix for a system having m states.

Since P_{ij} represents the probability of moving from state i to state j in one step, the sum of probabilities in every row is 1. Therefore, for any row i,

$$P_{i1} + P_{i2} + \cdots + P_{im} = 1 \qquad (3\text{-}1)$$

or in general form:

$$\sum_{j=1}^{m} P_{ij} = 1 \qquad \text{for every} \quad i \qquad (3\text{-}2)$$

The set of transition probabilities in any row represents all possibilities of going from one state in the present period to one of the m states in the next period.

Let us assume that all the possible states of a system $S_1, S_2, \ldots, S_m$ are discrete and consider them only at discrete points in time, which we can designate as $t = 0, 1, 2, \ldots$. The probability of being in a given state S_i at a given point in time is $q_i(t)$. For example, $q_2(4)$ indicates the probability that the system will be in state 2 at time 4. Considering that at any given point in time the system must be in one and only one of the possible states, we have

$$\sum_{i}^{m} q_i(t) = 1 \qquad \text{for} \quad t = 0, 1, 2, \ldots \qquad (3\text{-}3)$$

FIGURE 4. Transition matrix for system with m states.

		State				
	To From	S_1	$\cdots$	S_j	$\cdots$	S_m
	S_1	P_{11}	$\cdots$	P_{1j}	$\cdots$	P_{1m}
	$\vdots$	$\vdots$		$\vdots$		$\vdots$
State	S_i	P_{i1}	$\cdots$	P_{ij}	$\cdots$	P_{im}
	$\vdots$	$\vdots$		$\vdots$		$\vdots$
	S_m	P_{m1}		P_{mj}		P_{mm}

This probability is known as *transient* or *state probability*. If all the transient probabilities at time t are available to us, then we can calculate them at time $t + 1$ as follows:

$$q_j(t + 1) = \sum_i^m q_i(t)P_{ij} \quad \text{for} \quad j = 1, 2, \ldots, m$$

$$\text{and} \quad t = 0, 1, 2, \ldots \tag{3-4}$$

or

$$q_1(t + 1) = q_1(t)P_{11} + q_2(t)P_{21} + \cdots + q_m(t)P_{m1}$$

$$q_2(t + 1) = q_1(t)P_{12} + q_2(t)P_{22} + \cdots + q_m(t)P_{m2} \tag{3-5}$$

$$\vdots$$

$$q_m(t + 1) = q_1(t)P_{1m} + q_2(t)P_{2m} + \cdots + q_m(t)P_{mm}$$

Therefore, the probability of being in state S_j at time $t + 1$ is equal to the probability of being in state S_i at time t, multiplied by the transition probability P_{ij}, summed over all possible states S_i. This is the basic concept of a Markov process.

We can use matrix notation and rewrite the above equations in matrix form,

$$\mathbf{q}(t + 1) = \mathbf{q}(t)\mathbf{P} \tag{3-6}$$

where $\mathbf{q}(t)$ and $\mathbf{q}(t + 1)$ are vectors of transient probabilities at times t and $t + 1$, respectively, and $\mathbf{P}$ is the matrix of transition probabilities. If the state probabilities at time $t = 0$ is known, then we can compute the state probabilities at any time t by solving Eq. (3-6) successively:

$$\mathbf{q}(1) = \mathbf{q}(0)\mathbf{P}$$

$$\mathbf{q}(2) = \mathbf{q}(1)\mathbf{P}$$

$$\vdots \tag{3-7}$$

$$\mathbf{q}(t) = \mathbf{q}(t - 1)\mathbf{P}$$

$$\mathbf{q}(t + 1) = \mathbf{q}(t)\mathbf{P}$$

The above equations indicate that the following analysis can be made:

$$\mathbf{q}(1) = \mathbf{q}(0)\mathbf{P}$$

$$\mathbf{q}(2) = \mathbf{q}(1)\mathbf{P} \quad = (\mathbf{q}(0)\mathbf{P})\mathbf{P} \quad = \mathbf{q}(0)\mathbf{P}^2$$

$$\vdots \qquad \vdots \qquad \vdots \tag{3-8}$$

$$\mathbf{q}(t) = \mathbf{q}(t - 1)\mathbf{P} = (\mathbf{q}(0)\mathbf{P}^{t-1})\mathbf{P} = \mathbf{q}(0)\mathbf{P}^t$$

$$\mathbf{q}(t + 1) = \mathbf{q}(t)\mathbf{P} \quad = (\mathbf{q}(0)\mathbf{P}^t)\mathbf{P} \quad = \mathbf{q}(0)\mathbf{P}^{t+1}$$

Therefore by having the state probabilities at $t = 0$ and the transition matrix $\mathbf{P}$, we can determine the probabilities of outcomes at any time in the future by multiplying the probability vector $\mathbf{q}_i$ by some power of the transition matrix $\mathbf{P}$.

As an example, assume that the matrix of transition probabilities is known as

$$\mathbf{P} = \begin{bmatrix} .4 & .3 & .3 \\ .1 & .2 & .7 \\ .5 & .3 & .2 \end{bmatrix}$$

Suppose that the state probability vector at $t = 0$ is given as

$$\mathbf{q}(0) = [q_1(0) \quad q_2(0) \quad q_3(0)] = [.25 \quad .4 \quad .35]$$

We can find the state probabilities for a given time in the future, i.e., $\mathbf{q}(1)$, $\mathbf{q}(2)$, . . ., by successively solving Equation (3-6).

$$\mathbf{q}(1) = [q_1(0) \quad q_2(0) \quad q_3(0)] \times \begin{bmatrix} P_{11} & P_{12} & P_{13} \\ P_{21} & P_{22} & P_{23} \\ P_{31} & P_{32} & P_{33} \end{bmatrix}$$

$$= [q_1(1) \quad q_2(1) \quad q_3(1)]$$
$$q_1(1) = q_1(0)P_{11} + q_2(0)P_{21} + q_3(0)P_{31}$$
$$q_2(1) = q_1(0)P_{12} + q_2(0)P_{22} + q_3(0)P_{32}$$
$$q_3(1) = q_1(0)P_{13} + q_2(0)P_{23} + q_3(0)P_{33}$$

This procedure holds for the computation of $\mathbf{q}(2)$, $\mathbf{q}(3)$, ... By substituting the numbers given in our example for state probability and transition matrix, we have

$$\mathbf{q}(1) = [.25 \quad .4 \quad .35] \times \begin{bmatrix} .4 & .3 & .3 \\ .1 & .2 & .7 \\ .5 & .3 & .2 \end{bmatrix}$$

$$= [.315 \quad .26 \quad .425]$$

$$\mathbf{q}(2) = [.315 \quad .26 \quad .425] \times \begin{bmatrix} .4 & .3 & .3 \\ .1 & .2 & .7 \\ .5 & .3 & .2 \end{bmatrix}$$

$$= [.3645 \quad .274 \quad .3615]$$

$$\mathbf{q}(3) = [.3645 \quad .274 \quad .3615] \times \begin{bmatrix} .4 & .3 & .3 \\ .1 & .2 & .7 \\ .5 & .3 & .2 \end{bmatrix}$$

$$= [.35395 \quad .2726 \quad .37345]$$

$$\mathbf{q}(4) = [.35395 \quad .2726 \quad .37345] \times \begin{bmatrix} .4 & .3 & .3 \\ .1 & .2 & .7 \\ .5 & .3 & .2 \end{bmatrix}$$

$$= [.355565 \quad .27274 \quad .371695]$$

The important point to note here is that in the long run, as the number of steps t approaches infinity, the probability vector $q(t)$ approaches a constant value known as the *steady state* or *equilibrium*. This does not mean that process will stop in any one state, but that the probability of being in a given state approaches a constant value.

By introducing steady-state probabilities, the following formulation can present the steady-state probability vector q, where

$$q = qP \qquad (3\text{-}9)$$

In full, Eq. (3-9) can be presented as

$$[q_1 \quad q_2 \cdots q_m] = [q_1 \quad q_2 \cdots q_m] \times \begin{bmatrix} P_{11} & P_{12} & \cdots & P_{1m} \\ P_{21} & P_{22} & \cdots & P_{2m} \\ P_{m1} & P_{m2} & \cdots & P_{mm} \end{bmatrix} \qquad (3\text{-}10)$$

The matrix multiplication of Eq. (3-10) results in a system of m simultaneous linear equations that are not linearly independent. This happens because the sum of the elements in state probability vector equals unity:

$$\sum_{i=1}^{m} q_i = 1$$

Therefore, keeping this in mind, we can solve the system of simultaneous linear equations and derive a solution for the steady-state vector q. Returning to our example, we have

$$[q_1 \quad q_2 \quad q_3] = [q_1 \quad q_2 \quad q_3] \times \begin{bmatrix} .4 & .3 & .3 \\ .1 & .2 & .7 \\ .5 & .3 & .2 \end{bmatrix}$$

In order to find the steady state probabilities, we solve the system of simultaneous equations for q_1, q_2, and q_3. Thus,

$$q_1 = .4q_1 + .1q_2 + .5q_3$$

$$q_2 = .3q_1 + .2q_2 + .3q_3$$

$$q_3 = .3q_1 + .7q_2 + .2q_3$$

We know that $q_1 + q_2 + q_3 = 1$, so solving the system we have

$$q_1 = .35538, \qquad q_2 = .27272, \qquad q_3 = .37190$$

It is interesting to observe what happens when we power the transition matrix P in the above example:

$$P^2 = \begin{bmatrix} .4 & .3 & .3 \\ .1 & .2 & .7 \\ .5 & .3 & .2 \end{bmatrix} \times \begin{bmatrix} .4 & .3 & .3 \\ .1 & .2 & .7 \\ .5 & .3 & .2 \end{bmatrix} = \begin{bmatrix} .34 & .27 & .39 \\ .41 & .28 & .31 \\ .33 & .27 & .40 \end{bmatrix}$$

$$P^4 = \begin{bmatrix} .34 & .27 & .39 \\ .41 & .28 & .31 \\ .33 & .27 & .40 \end{bmatrix} \times \begin{bmatrix} .34 & .27 & .39 \\ .41 & .28 & .31 \\ .33 & .27 & .40 \end{bmatrix} = \begin{bmatrix} .355 & .2727 & .3723 \\ .3565 & .2728 & .3707 \\ .3549 & .2727 & .3724 \end{bmatrix}$$

If we carry on the operation, before long we get the following matrix:

$$\begin{bmatrix} .35538 & .27272 & .37190 \\ .35538 & .27272 & .37190 \\ .35538 & .27272 & .37190 \end{bmatrix}$$

It is interesting to note that each row in the matrix becomes the same as the steady-state probability vector.

It should be noted, however, that in some systems it takes a long time to obtain the steady-state vector, and in others it is impossible to approach the steady state in the system.

Markov chain analysis can be used in any decision-making situation that requires the prediction of the system's behavior. The more recent application of Markov chains has been in the area of policy determination. Specially developed algorithms permit the selection of optimal policies for such problems as improvement of productivity, machine replacement, and equipment maintenance.

IV. DECISION UNDER UNCERTAINTY

This situation exists when the decision maker is unable to assign probabilities to the likelihood of occurrences of future events. When confronted with this situation, there are a number of different criteria that may be used; however, it is difficult to say which is best, the choice depending on the decision maker.

A. The Criterion of Optimism (Maximax or Minimin)

This criterion assumes that the decision maker is completely optimistic and will select the alternative with the largest profit or payoff. Therefore, the decision maker will choose the largest payoff under each contingency. By so doing, the payoff matrix is reduced to a column vector, and the next step is to choose the remaining alternative with the highest payoff.

For example, suppose that the management of a fast-food restaurant is contemplating construction of a new restaurant. Three location sites are under consideration, but their relative attractiveness depends upon the rate of population growth over the next five years in each location.

TABLE 1. The Payoff Matrix: Anticipated Profit for the Next Five Years

	Contingencies: Population growth rate			
Alternatives	No growth	Low	Medium	High
Location A	− 150,000	300,000	600,000	1,000,000
Location B	− 100,000	400,000	500,000	800,000
Location C	− 200,000	− 100,000	700,000	900,000

Four possible growth rates and the anticipated profit for the next five years are given in Table 1. If the probability of the future growth rates are unknown, which location should be selected?

Under the *maximax* principle, the management would select the highest possible outcome for each location. This is summarized as follows:

Location	Best possible outcome
A	1,000,000
B	800,000
C	900,000

The location with the highest return (A) is chosen.

If we are dealing with cost rather than profit, the optimistic decision maker would use the criterion of *minimin*. In this case the decision maker would first choose the lowest cost of each alternative and then choose the lowest possible cost among those.

B. The Criterion of Pessimism (Maximin or Minimax)

A pessimistic decision maker would tend to choose this criterion. Under the *maximin* principle, the worst possible outcome for each alternative would be sought and the best of the worst would be chosen.

In the site selection example, the management would choose the worst outcome for each alternative. For location A, the worst that can happen is a loss of $150,000 when no growth is experienced. Similarly, the worst possible outcome for B is a loss of $100,000, and for C, a loss of $200,000. Having done this, the next step is to choose the best of the worst possible outcomes, namely, location B with a loss of $100,000 under the no-growth contingency.

If the pessimistic decision maker is minimizing cost rather than maximizing profit, the criterion of *minimax* will be used. In this case the highest cost under each contingency for every alternative is selected and the alternative with the least cost is accepted.

C. Maxim or Minim (Hurwicz Criterion)

Under the criterion of *maxim* or *minim*, the decision maker is neither completely optimistic nor completely pessimistic. Therefore, it is necessary to decide the decision maker's degree of optimism or pessimism on a 0–1 scale. For example, if the decision maker is 60% optimistic about an alternative, this indicates a 40% pessimism about the same alternative.

Under the maxim criterion, the smallest value of an alternative under any contingency is selected and multiplied by the degree of pessimism of the decision maker. This result is then added to the result of the multiplication of the largest value of the same alternative by the degree of optimism of the decision maker. This is called the *weighted value* of the alternative.

Following the same procedure, the weighted values of all alternatives are computed and the alternative with the highest weighted value is selected.

Let us consider the site selection example. If the management is 60% optimistic and 40% pessimistic, then using the figures in Table 1, we have the following:

Location	Highest possible outcome	Lowest possible outcome
A	1,000,000	−150,000
B	800,000	−100,000
C	900,000	−200,000

The weighted value of each alternative can be determined as follows:

Location	Weighted value
A	.6 × 1,000,000 + .4 × (−150,000) = $540,000
B	.6 × 800,000 + .4 × (−100,000) = $440,000
C	.6 × 900,000 + .4 × (−200,000) = $460,000

Examining the weighted value of each alternative, location A with the highest weighted value is selected.

For cost minimization, the weighted value of an alternative is computed by multiplying the highest cost of the alternative under any contingency by the degree of optimism of the decision maker (*minim*).

The only difficulty in applying this criterion, obviously, is determining the degree of optimism or pessimism of the decision maker.

D. The Criterion of Insufficient Reason (Laplace)

This criterion assumes that since we know nothing about the probabilities of any contingency, we assume that they are all equally likely to occur. After assigning equal probabilities to occurrence of each contingency, the *expected value* of each alternative is computed. The one with the highest (in case of profit maximization) or lowest (in case of cost minimization) expected value is selected.

Using the example given previously, the expected values for location sites are as tabulated:

Location	Expected value
A	¼(−150,000) + ¼(300,000) + ¼(600,000) + ¼(1,000,000) = $437,500
B	¼(−100,000) + ¼(400,000) + ¼(500,000) + ¼(800,000) = $400,000
C	¼(−200,000) + ¼(−100,000) + ¼(700,000) + ¼(900,000) = $325,000

Therefore, the best location is *A* with an expected profit of \$437,500. The only shortcoming of this criterion is the assignment of equal probabilities when there is no reason, save intuitive appeal, to make such an assumption.

V. DECISION UNDER CONFLICT

Decision under conflict occurs when the states of nature facing the decision maker are the potential courses of action for a competitor. In situations such as these, *game theory* is applicable. The primary objective of game theory is the development of a rational criterion for selecting a strategy or plan of action. It requires that the outcomes of various strategies of the participants be expressed as numerical values.

In competitive decision making, two or more decision makers, considered to be equally informed and intelligent, are placed in contest against each other. They are referred to as *players,* and their conflicts are called *games.* The rationale of their competition is the basis of game theory.

The first step in analysis under game theory is to examine the possible outcomes in the game. It is assumed that these outcomes are well specified in advance and that each player in a game has a consistent pattern of preferences among the outcomes.

In a game context, alternative courses of action are called *strategies.* There is a slight difference between alternatives in games of skill, chance, and strategy. In the last category, the best course of action for a player depends on the opponents' alternatives. Thus an optimal strategy in a competitive environment may be always to use one alternative or to mix alternatives for successive plays.

The simplest example of game theory is the two-person game. The situation is represented by a matrix with one player's alternatives on the side and the opponent's alternatives across the top. Multiplayer games rely on essentially the same type of inductive and deductive reasoning as do two-person games, but with significantly increased mathematical complication without a corresponding accrual of basic logic. Actually, the two-player limitation is not as restrictive as it might seem. Strategy for one side in a conflict is often designed to counter the aims of the single most dangerous member of an opposing team. In other cases a player may view the entire array of business competitors as a single opponent and thereby reduce the conflict to a two-person contest.

In two-person games the rows of the payoff matrix contain the outcomes for one player and the columns show the outcomes for the other player. Let us assume that individual *A* plays a game with another individual, *B.* The alternatives for player *A* are A_1, A_2, A_3, and A_4. Player *B* also has four possible courses of action: B_1, B_2, B_3, and B_4. Table 2 shows the payoff to player *A* for each combination of alternatives. A positive number

TABLE 2. Payoff Matrix for a Two-Person Game

Player A strategies	Player B strategies			
	B_1	B_2	B_3	B_4
A_1	-2	-3	5	6
A_2	-1	3	-4	-3
A_3	2	4	7	3
A_4	-3	-2	4	1

is always a gain for A and a negative number is a gain for B. We consider a negative payoff as loss and a positive payoff as a gain. If player A chooses alternative A_1 and B selects B_1, the outcome is a loss of \$2 for player A and a gain of \$2 for B (i.e., A pays B \$2). At A_3, B_3, player B pays A \$7.

The payoff to player A is always the negative of that for player B. Therefore, the sum of the payoffs for any choice of alternatives is always zero:

$$\text{for} \quad (A_1, B_1), \quad (-2) + (+2) = 0$$

$$\text{for} \quad (A_3, B_3), \quad (+7) + (-7) = 0$$

The first step in analyzing a game matrix is to check for *dominance,* a condition in which one alternative will always be better for a player than another (e.g., A_3 dominates A_4). In competitive decisions the columns as well as the row represent alternatives, so it is necessary to check both horizontally and vertically. Furthermore, the discovery of one dominant relationship may reveal another dominant condition that otherwise would not be recognized.

A sequential dominance relationship can be observed in Table 2. A check of the columns, the alternatives available to player B, indicates that no one alternative is always better than another for B whatever alternative A chooses. For instance, B_3 is preferred to B_4 for the first two rows, but not the second two.

The assumption that allows two-way dominance checks is that both players are intelligent. Thus B would recognize that A would never use A_4 (since A_3 is dominant), which means B would never use B_4. Therefore, alternatives A_4 and B_4 (i.e., fourth row and fourth column in the payoff matrix shown in Table 2) could be eliminated, yielding the payoff matrix shown in Table 3.

Using the same alternative every time is called a *pure strategy*. If both players use pure strategies, we say a *saddle point* is present. A saddle point is recognized by an outcome that is both the smallest number in its row and the largest number in its column (or vice versa). The outcome of 2 at (A_3, B_1) satisfies the requirements for a saddle point.

TABLE 3. Payoff Matrix for a Two-Person Zero-Sum Game

Player A strategies	Player B strategies		
	B_1	B_2	B_3
A_1	-2	-3	5
A_2	-1	3	-4
A_3	2	4	7

The significance of a saddle point develops from an investigation of the player's motives. Alternatives A_1 and A_2 are attractive to player A since they allow potential gains of $5 and $3. However, B can be as sure that A will lose $3 or $1 by using B_2 or B_1, respectively, whenever A chooses strategies A_1 or A_2. A would be most attracted to A_3 because no negative outcomes can appear from this alternative. Player B would be attracted to B_1 since this strategy would guarantee a loss no worse than $2. Player A can observe the advantage B_1 affords B, and would select alternative A_3 to maximize overall payoff. Therefore both players would use pure strategy with A always employing alternative A_3 and B always using B_1.

The underlying principle of choice for the players is maximin–minimax. B seeks the lowest maximum loss (minimax); A identifies the highest minimum gain (maximin). This assumption explains the fact that each player is capable of selecting the strategy that will maximize gain or minimize loss.

When the optimal strategy for both players is a pure strategy, the *value of the game* is the outcome at the saddle point. In our example, the value of the game is $2.

When no saddle point exists, the players turn to a policy called *mixed strategy*. This means that the alternative employed for each play is a random choice from those available. The value of the game is the average return resulting from each player's having followed the optimal mixed strategy.

As is often the case with complex mathematical models, most actual games are extremely complicated to analyze in detail. However, the result of this technique, as with many others, is manifold. It can be a useful aid in delineating the various alternative courses of action open to management. This technique holds promise for increased use through the application of digital computer techniques.

This chapter has highlighted the role of decision theory in assisting management to solve complex decision problems. The choice of criteria depends upon the type of problem under consideration.

It is more difficult to deal with decision under uncertainty than under certainty or risk. Uncertainty should be avoided in decision making and efforts should be made to convert it to a form of risk or certainty. Most management science tools and techniques solve problems that fall within these two categories.

It should be noted that management science techniques do not dictate the final decision. Real-world problems are so complex that a combination of experience, judgment, and analytical tools is often necessary to find an adequate solution for them.

REVIEW QUESTIONS

1. What is the difference between making a decision and solving a problem?

2. Contrast decisions under risk and those under uncertainty.

3. What is subjective probability and how does it differ from objective probability?

4. Explain under what circumstances Bayes' theorem should be used.

5. List all the factors involved in constructing a decision tree.

6. Present a general form of a transition matrix with five states.

7. Define transient or state probability.

8. Discuss the nature of different criteria available to management for decisions under uncertainty.

9. What is meant by a saddle point in game theory? Explain its significance.

10. What is the underlying principle of choice for players in game theory?

RECOMMENDATIONS FOR FURTHER READING

Ahlers, David M., An Investment Decision Making System, *Interfaces* 5 (2), 72–90 (February 1975).

Berlin, Victor N., Administrative Experimentation: A Methodology for More Rigorous "Muddling Through," *Management Science* 24 (8), 789–799 (April 1978).

Bierman, H., Jr., Bonini, C. P., and Housman, W. H., *Quantitative Analysis for Business Decisions*. Homewood, Illinois: Richard D. Irwin, 1977, 5th ed.

Burt, John M., Planning and Dynamic Control of Projects Under Uncertainty, *Management Science* 24 (3), 249–258 (November 1977).

Cyert, R. M., DeGroot, M. H., and Holt, C. A., Sequential Investment Decisions with Bayesian Learning, *Management Science* 24 (7), 712–718 (March 1978).

Derman, C., Lieberman, G. J., and Ross S. M., A Renewal Decision Problem, *Management Science* 24 (5), 554–561 (January 1978).

Eliashberg, Jehoshus, and Winkler, Robert L., The Role of Attitude Toward Risk in Strictly Competitive Decision-Making Situations, *Management Science* 24 (12), 1231–1241 (August 1978).

Harison, E. F., *The Managerial Decision-Making Process*. Boston: Houghton Mifflin, 1975.

Kaufman, A., *Methods and Models of Operations Research*. Englewood Cliffs, New Jersey: Prentice-Hall, 1963.

Konczal, Edward F., Documenting Large-Scale Telecommunications Computer Analyses, *Journal of Systems Management* 29 (6), 14–17 (June 1978).

Miller, D. W., and Starr, M. K., *Executive Decisions and Operations Research*, Englewood Cliffs, New Jersey: Prentice-Hall, 1960.

Montanari, John R., Managerial Discretion; An Expanded Model of Organizational Choice, *Academy of Management Review* 3 (2), 231–241 (April 1978).

Olson, Philip D., Notes—Decision-Making Type 1 and Type 2 Error Analysis, *California Management Review* 20 (1), 81–83 (Fall 1977).

Oppenheimer, Kenneth R., A Proxy Approach to Multi-Attribute Decision Making, *Management Science* 24 (6), 675–689 (February 1978).

Raiffa, H., *Decision Analysis*. Reading, Massachusetts: Addison-Wesley, 1970.

Schwarz, Leroy B., and Johnson, Robert E., An Appraisal of the Empirical Performance of the Linear Decision Rule for Aggregate Planning, *Management Science* 24 (8), 844–849 (April 1978).

Thomas, H., *Decision Theory and the Manager*. London: Pitman, 1972.

Vazsony, Andrew, Information Systems in Management Science–Decision Support Systems: The New Technology of Decision Making, *Interfaces* (*TIMS*) 9 (1), 72–77 (November 1978).

White, D. J., *Decision Methodology*. London: Wiley, 1975.

Chapter 4
Quantitative Methods for Management

INTRODUCTION

Quantitative management science tools and techniques rely on mathematical and statistical methods to provide either solutions or more efficient analyses of managerial problems. The reader is not likely to be an expert in probability and statistics; still, the management scientist should be able to communicate with those specialized in the use of these mathematical tools.

The purpose of this chapter is to recommend the basic mathematical, probabilistic, and statistical tools needed for management science work. A survey comprised of three general questions served as a guide for reaching these recommendations. The questions were as follows:

1. What quantitative mathematical techniques are required for systems and management science work?
2. Is too much mathematics being required from students preparing for work in management science and systems engineering?
3. How much mathematics is generally used in industry in management science and systems work?

This chapter compares the quantitative tools taught in the universities with those required by industry for the problem-solving and decision-making processes. There is a gap between the two that should be bridged. This chapter also covers the quantitative methods with which the management scientist must be familiar in order to be able to work with specialists. The Appendix of the chapter gives a summary relating the mathematical, probabilistic, and statistical tools to the various management science techniques.

I. QUANTITATIVE METHODS REQUIREMENTS AT UNIVERSITIES

The mathematics requirements for students in management science and systems are continuously reviewed and updated by universities. Such issues as the distinction between pure mathematics and applied mathematics are being discussed, with applied mathematics receiving increasing attention as an aid in solving problems that arise in the real world. Certainly, educators in the business schools are taking a closer look at their mathematical needs. Mathematics requirements in several undergraduate business curricula are being reexamined. Concentration on quantitative methods is needed in order to achieve competence in management science.

In the early seventies, educators believed that business students needed applied mathematics to solve problems in the areas of decision theory, risk analysis, and operations research. At a minimum, students of business would have to know enough of the language of mathematics to be able to use that which is relevant to their problems. At that time, most business schools offered a core curriculum in mathematics that would allow the student to read management science material and lay a foundation for further elective study in the quantitative methods area. This curriculum included calculus, probability and statistics, matrix algebra, and some work with an electronic computer in a mathematical context. The core was taught with management science applications in mind, but such applications were introduced sparingly and only for purposes of motivation, the emphasis being on mathematics. Applications per se were treated elsewhere, either as courses or as parts of courses. Students going on for graduate study were offered a more intensive mathematical curriculum that included differential equations, calculus of several variables, and mathematical computation on an electronic computer.

It is interesting to note how business schools are now upgrading and increasing the math requirements of their students. Several educators insist that a new approach is needed in the study of systems and management science in business schools. In the past, teaching in business schools concentrated on the management-by-practice approach. This consisted of emphasizing the various functional business areas of marketing, production, finance, and personnel. The basic course on principles of management and business policy centered on the study and application of the statements of practitioners and armchair theorists. Recent contributions of research methodology and management science have made a large impact on today's managers. The methods of solving business problems have grown rapidly. Management science now applies analysis and synthesis to the development of systems models for the solution of business problems. As a result of the optimization techniques of management science, the business firm has come to be regarded as a system rather than a collection of isolated problems.

Modern business needs essentially two kinds of systems people: One is the systems manager who functions as a gross systems designer and decision maker and is responsible for solving major problems, making major decisions, and providing direction to the systems specialists and to those who perform the operational tasks. The second kind of individual is the systems designer, who applies a high degree of expertise to the translation of performance specifications and gross system design into an operational reality. The systems approach necessitates a complete break from the old functional approach to the study of business.

Today, educators in business schools differentiate the needs of modern business into two categories: managers and designers, and they offer their curricula accordingly. Most business schools offer courses in the following areas related to the quantitative methods needs by the systems designer:

Computer science

Probability and statistics

Finite math with business applications

Calculus with business applications

Modeling and simulation

The topics covered under computer science include the basic components of computers, computer logic, and the capabilities and limitations of computers. Probability and statistics are recommended and, indeed, are presently included in many present courses. Finite mathematics with business applications encompasses a wide variety of topics related to a plausibility and application perspective. Calculus with business applications includes selected topics in calculus developed from a similar perspective. Topics related to a combined application of quantitative techniques and computer science to business systems problems fall under modeling and simulation.

These courses are offered at different levels for undergraduate and graduate students, depending on the student's degree of understanding, knowledge, skill in application and analysis, and the evaluation that the student anticipates in postgraduate work.

According to some experts on the art of modeling, skill in modeling involves a sensitive and selective perception of management situations. This leads the management scientist to search for analogies or associations with already well developed logical structures.

Other sources of modeling skill is the study of mathematics for management science work. It is obvious that a feeling of being at ease with mathematics is important. One studies advanced mathematics, despite the fact that it will probably not be "useful," to achieve a relaxed grasp of the less advanced mathematics that probably will. This one idea has probably been partially responsible for the extensive study of advanced mathematics by many students in management science. Is this extensive mathematics study worthwhile? Perhaps a look at some of the mathematics used by industry in management science and systems work will give some insight into the problem.

II. QUANTITATIVE REQUIREMENTS IN INDUSTRY

Quantitative analysis has been increasingly used since World War II in industrial and commercial operations. This increase has necessitated a complete rethinking of many industrial concepts of management. Two apparent trends emerge from this swing to quantified methodology:

1. The development of a new hierarchy of analytical tools distinct from all concepts of quantitative analysis used heretofore in business and industry
2. The utilization of electronic computers to identify the complex dynamic processes of business and industry in order to describe their behavior and control them.

Two forces that compel the adoption of the scientific method in business administration are the continuing rapid expansion of the economy and the advent of computers. No sooner did electronic computers appear on the scene than they were adapted to commercial data processing. Their use created demands for new skills: computer programming (a facility in the language and grammar of computing) and systems analysis (the elaboration of programming to suit the wider functions of management decision making and conceptualization). These new skills called for a higher level of abstract reasoning and interest in mathematics than was normally required in conventional administrative functions.

About the time computers first appeared in commercial and industrial operations, quantitative techniques were beginning to be employed in dealing with the problems of business management. Many of these techniques have come to be grouped under the heading of *decision theory*. They are usually presented in mathematical or logical terms as mathematical models. Operations research applications, being mostly quantitative, increased the importance of mathematics in business administration. While computers have become the "instruments of the hand" of the quantitative techniques of decision theory, mathematics has provided business administration with the instruments of the mind.

It is worthwhile to look briefly at the thought effectiveness that quantitative analysis has historically brought to management. In the first phase of the development of quantitative techniques, arithmetic tools were used to search for answers to problems in commerce, finance, and industry. In the second phase, differential analysis, which emerged from Newton's principles of infinitesimal calculus, was developed and adapted to economics and mathematical theories such as marginal utility. Next came the development of the analysis of stochastic events—the awareness that a range of stochastic values can be gathered under a single quantitative index of mathematical expectation. The fourth phase of quantification in business administration was introduced in the form of a game-theory approach to business. According to some views, the applications of the concepts of decision theory are apparent in operations research.

Mathematics has attained significance in industrial management science because of a growing need for abstract thinking and formal reasoning. This need has been attributed to such things as the growing use of computers, the increasing applications of operations research, and the great numbers of scientific people entering the field of management science.

It is important to recognize the high value industry has recently placed on advanced education. Also, technological advances have placed professional people in particular demand. Today, one of the strongest determinants of site location is accessibility to at least one high-quality university or technical institute. Professional and technical personnel demand top quality community educational facilities. Since quality educational opportunities at all levels must be available in communities competing for new industries, there is little doubt that industries will be demanding excellence in the management science curriculum. This curriculum will have to include the mathematics required by industry for its management science and systems work.

Industrial management science uses both analytical and simulation methods. Mathematical analysis consists of two stages. First, an observed real system must be cast into the form of a mathematical model. The correspondence between reality and the meaning attached to the components of the model is crucial. Once the model has been formulated, the use of deductive mathematical methods yields a solution to the problem. This solution is guaranteed to be a correct interpretation of the model. Of course, while it may be correct for the model, it may be false in terms of reality if the original model–reality correspondence was inappropriate.

Simulation methods, like analytical ones, have to be concerned with the correspondence between reality and a model. But even if this correspondence is appropriate, simulation methods do not provide guarantees that a correct solution will be derived. Making observations in the form of inferences will not guarantee a correct solution. At most we shall have a probabilistic estimate of their correctness. Analytical solution techniques are generally preferred. The reason that we employ simulation methods is that we do not have adequate analytic techniques for certain models.

The convenience of mathematical analysis in systems management comes from the fact that the solution procedures are derived within the formal mathematical system and apply regardless of the meaning attached to the variables of a particular model. With this in mind, it might be useful to consider the three basic elements of any mathematical structure:

1. *Postulates, axioms, or assumptions.* These are a set of propositions that are not derivatives of other propositions. Euclid considered them self-evident truths. It was not until the 19th century that the broader characteristic of man-made assumptions replaced the notion of self-evident truths.

2. *Logic.* The human cognitive process which can manipulate the assumptions or postulates according to the rules of rational operations. This is a deductive process.

3. *Derived propositions.* New, derived propositions usually referred to as theorems, yielded by the application of logic to the basic assumptions or postulates.

All the discussions mentioned above suggest that the basic contribution of the systems designer is the application of special knowledge to the solution of systems problems. Therefore training should accentuate greater technical skills and knowledge of business problems, MIS systems, management science, and computer applications. The systems designer must be able to formulate decision rules, use simulation techniques, and apply the principles of modeling and other techniques. The ideal systems designer will have an undergraduate degree in either operations research, mathematics, or management science. Those lacking such a background should have the potential to be trained in quantitative methods.

There are two main types of business problem where quantitative techniques may be applicable. One is where several courses of action are possible and each is evaluated to determine the best one. The other category includes problems in which currently available alternative techniques are no longer suitable and a new approach must be developed.

The other techniques with which the systems person should be familiar include those related to statistical and probability theory. Statistical techniques can be applied to a population area by some form of random sampling. From these samples certain inferences may be made. At given levels of confidence the predictions are representative of the population being analyzed.

Systems analysis can be defined as a quantified decision-making procedure, as just one term for an approach to problem solving that good management has always practiced. Quantification must start in the development phase of the life cycle of the system or equipment, during the period when technical performance characteristics are being developed. In order to control development, we must quantify objectives, identify necessary inputs, measure progress, and verify outputs.

Recent trends in education related to the systems field, emphasize management information systems. Some of the training programs needed by those in business obviously do not meet the educational goals of university students. A business must look for shorter-range returns for its large investment in training; it cannot afford to provide long periods of education whose payoff may be both intangible and remote in time. Therefore, companies are mainly concerned with bringing their personnel up to date with new technologies so that the individuals may continue to remain abreast on their own initiative. This applies to training programs for management information systems as well as any other programs.

Leaders in industry and government are groping for ways to take advantage of the tremendous payoffs possible with computer-based information systems. The change from straight data processing to computer-based-systems problem solving has been fast and its potential in decision making has stimulated enthusiasm among government and business leaders.

The above discussion gives some indication of what kind of mathematics background industry wants its management scientists to have. Before drawing some conclusions, the industrial–academic interface will be briefly considered.

III. THE GAP BETWEEN INDUSTRY AND ACADEMIA

There is a gap between what industry wants and what the universities are providing. Many believe that the traditional concept of industrial–academic interaction has been and continues to be superficial and inadequate. Three components are involved in this interaction:

1. The contribution of money by the industries, in the form of gifts or taxes, to help support educational institutions
2. The transfer of students from academia to the industrial world
3. Some exchange of substantive information acquired in research laboratories.

Direct industrial support of academic institutions is not very impressive because during the past two decades there has been a steady growth of public financing. Students, the principal product of universities, are being produced in large numbers. Many of them are bright and well-informed.

Still, industry is not entirely enchanted with them since many of the brightest are not interested in industrial careers or in using their knowledge and creativity in the service of their employers.

The exchange of information between the universities and industry is also unsatisfactory. Information is accumulating at a rate that far exceeds the capacity of the human mind to integrate it. The potential of the computer to bridge the information-transfer gap may go a long way toward solving the problem, but there is another component of the information exchange problem: the necessary and desirable difference between the character and objectives of academic and industrial research. Most academic research is basically dedicated to finding general solutions to problems. In industrial research the primary objective is to obtain an answer to the problem at hand. Consequently, the methodologies of the two kinds of research must differ.

The really meaningful value of industrial–academic interaction comes from the fact that academic research does develop theories and models that have some general application. They can contribute to applied research in two ways. First, some of the models directly suggest solutions to applied problems. Second, and more important, academic research helps establish that the behavior of things in the universe is systematic.

Academia and industry have much in common. Certainly the amount and level of mathematics required for management science should be an important area of interaction between industry and the universities. Perhaps the universities should do some hard listening to industry's expressions of need for mathematics in this field.

IV. SUMMARY

There has been a great increase in the use of quantitative methods since World War II. New quantitative analysis tools and electronic computers are being used extensively in management science. The mathematical and logical skills of decision theory are being applied in industrial management science, and there is a growing need for abstract thinking and formal reasoning. Mathematics provides the language for this thinking and reasoning. Industrial mathematics requirements in management science would seem to include the applied mathematics of modeling, probability, statistics, quantified decision making, algebra, calculus, and matrices. Universities should thoroughly investigate these requirements to determine the levels of training that are necessary.

The author has accumulated a list of quantitative methods (Appendix to this chapter) that he considers essential to enable the management scientist to identify the mathematical, probabilistic, and statistical tools available for problem solving. The level of competence needed in handling these quantitative tools depends on whether they will be applied in optimization, simulation, or control.

REVIEW QUESTIONS

1. What types of mathematics should be considered for management science schools? For business schools?

2. Give examples of industrial problems that can be solved by management science techniques.

3. Explain the phases of the development of quantitative techniques application in business administration.

4. What kind of educational training should an individual have to become a systems designer?

5. Explain why mathematics has attained significance in industrial management science.

6. What is the basic contribution made by a systems designer when designing a system?

7. Give examples of business problems that are best solved by the use of quantitative analysis.

8. What kind of thinking is required in systems analysis?

9. Explain the existing gap between what industry needs and what the universities are providing.

RECOMMENDATIONS FOR FURTHER READING

Text Books

Cook, T. A., and Russell, R. A., *Introduction to Management Science*. Englewood Cliffs, New Jersey: Prentice-Hall, 1977.

Eppen, Gary D., and Gould, F. J., *Quantitative Concepts for Management*. Englewood Cliffs, New Jersey: Prentice-Hall, 1979.

Freund, John E., *Statistics*. Englewood Cliffs, New Jersey: Prentice-Hall, 1970.

Gaver, D. P., and Thompson, G. L., *Programming and Probability Models in Operations Research*. Monterey, California: Brooks/Cole, 1973.

Johnson, Rodney D., and Siskin, Bernard R., *Quantitative Techniques for Business Decisions*. Englewood Cliffs, New Jersey: Prentice-Hall, 1976.

Kemeny, J. G., Schleifer, Jr., A., Snell, J. C., and Thompson, G. L., *Finite Mathematics with Business Applications*. Englewood Cliffs, New Jersey: Prentice-Hall, 1972, 2nd ed.

Lazarus, M., The Elegance and the Relevance of Mathematics, *The Chronical of Higher Education* XI (12), (1 December 1975).

Levin, Richard I., and Kirkpatrick, Charles A., *Quantitative Approaches to Management*. New York: McGraw Hill, 3rd Edition, 1975.

Ross, Sheldon M., *Applied Probability Models with Optimization Applications*. San Francisco: Holden-Day, 1970.

Shore, Barry, *Quantitative Methods for Business Decisions—Text and Cases*. New York: McGraw Hill, 1978.

Articles

Aumann, R. J., Some Thoughts on the Minimax Principle, *Management Science* 18 (5), 54 (January 1972, II).

Deb, Rajat K., Optimal Dispatching of a Finite Capacity Shuttle, *Management Science* 24 (13), 1362–1372 (September 1978).

Dyer, James S., A Time-Sharing Computer Program for the Solution of the Multiple Criteria Problem, *Management Science* 19 (12), 1379–1383 (August 1973).

Graves, Stephen C., and Haessler, Robert W., On "Production Runs for Multiple Products: The Two-Product Heuristic," *Management Science* 24 (11), 1194–1196 (July 1978).

Hallbauer, Rosalie C., Statistics and Management Science—Their Impact on Standard Costs, *RIA Cost and Management* 49 (3), 18–27 (May–June 1975).

Halpern, Jonathan, Finding Minimal Center-Median Convex Combination (Centdian) of a Graph, *Management Science* 24 (5), 535–544 (January 1978).

Keeney, Ralph, Utility Functions for Multiattributed Consequences, *Management Science* 18 (5), 276 (January 1972, I).

Lageweg, B. J., Lenstra, J. K., and Rinnooy-Kan, A. H. G., Job-Shop Scheduling by Implicit Enumeration, *Management Science* 24 (4), 441–450 (December 1977).

Lemoine, Austin J., Network of Queues—A Survey of Weak Convergence Results, *Management Science* 24 (11), 1175–1193 (July 1978).

Rosenthal, Richard E., White, John A., and Young, Donovan, Stochastic Dynamic Location Analysis, *Management Science* 24 (6), 645–653 (February 1978).

Saniga, Erwin M., Joint Economically Optimal Design of X and R Control Charts, *Management Science* 24 (4), 420–431 (December 1977).

Schoner, Bertram, and Mann, J. K., Probability Revision Under Act—Conditional States, *Management Science* 24 (15), 1650–1657 (November 1978).

Stidham, Shaler, Jr., Socially and Individually Optimal Control of Arrivals to a GI/M/1 Queue, *Management Science* 24 (15), 1598–1610 (November 1978).

Sweeney, Dennis J., Winkofsky, E. P., and Roy, Probir, et al., Composition vs. Decomposition: Two Approaches to Modeling Organizational Decision Processes, *Management Science* 24 (14), 1491–1499 (October 1978).

Ward, J. B., Determining Reorder Points When Demand Is Lumpy, *Management Science* 24 (6), 623–632 (February 1978).

APPENDIX TO CHAPTER 4
QUANTITATIVE METHODS REQUIRED
FOR MANAGEMENT SCIENCE

I. Mathematics

    A. *Sequences*
        1. Arithmetic Progressions
        2. Geometric Progressions
        3. Weighted Moving Averages
        4. Exponential Smoothing

    B. *Relations, Functions, and Graphs*
        1. Linear Functions
            a. Slope
            b. Sectional Continuity
        2. Quadratic Functions
        3. Exponential Functions
        4. Logarithmic Functions
        5. Line of Regression

    C. *Matrices*
        1. Column and Row Vectors
        2. Operations on Matrices
            a. Addition and Subtraction
            b. Multiplication of Matrices
        3. Definitions of Terms
            a. Identity or Unit Matrix
            b. Matrix Inverse
            c. Matrix Transpose

    D. *Linear Programming*
        1. Recognition of the Formulation of Linear Programming Problems
        2. Simplex Method
            a. Nonnegative Assumption
            b. Nondegeneracy Assumption
            c. Slack Variables
            d. Artificial Variables
            e. Basic Feasible Solution
        3. Concept of the Dual Linear Programming Problem
        4. Special Cases
            a. Transportation Problems
            b. Zero-Sum Games

    E. *Statements and Set Theory*
        1. Statements
            a. Atomic
            b. Compound
            c. Truth Tables
            d. Tree Diagrams

      2. Operating with Sets
        a. Complementation
        b. Union
        c. Intersection
        d. Venn Diagrams
      3. Critical Path Analysis
        a. Slack Time
        b. Free Slack
        c. Independent Slack

F. *Mathematics of Finance and Accounting*
      1. Compound Interest Problems
        a. Solve for Future Amount
        b. Solve for Present Value
        c. Solve for Interest Rate
        d. Solve for Number of Payments
      2. Depreciation Calculations
        a. Straight Line
        b. Sum-of-the-Years' Digits
        c. Double Declining Balance
      3. Break-Even Analysis
      4. Double Classification Bookkeeping (Distinguished from Double Entry)

G. *Computation Topics*
      1. Representation of Numbers
        a. Binary, Octal, Decimal, and Hexadecimal
        b. Floating Point
      2. Significant Digits
      3. Rounding
        a. Decimal Rounding with Binary Numbers
        b. Balancing Sums of Rounded Numbers
      4. Iteration
      5. Calculation as an Alternative to Table Look-up

H. *Counting or Enumeration*
      1. Permutations
      2. Combinations
      3. Compositions
      4. Partitions

II. Statistics and Probability

A. *Introduction to Statistics*
      1. Effective Uses of Statistics
      2. Misuses of Statistics
      3. Decision Parameters
        a. Frequency Distribution
        b. Averages
        c. Measures of Variation
        d. Measures of Relationship

B. *Elementary Probability Theory for Finite Sample Spaces*
   1. Probability
      a. Experiments
      b. Sample Spaces
      c. Events and Sets
      d. Mutually Exclusive and Independent Events
      e. Randomness
   2. Conditional Probability

C. *Introduction to Random Variables, Distributions, and Distribution Properties*
   1. Random Variables
      a. Random Variables and Their Probability Functions
      b. Mathematic Expectation of a Random Variable
      c. Mean, Average, and Variance of a Random Variable Function
      d. Probability Distribution
      e. Frequency Distribution
   2. The Normal Distribution
      a. Joint Probability Function of Two Random Variables
      b. Probability Graphs for Continuous Random Variables
      c. Probabilities Represented by Areas
      d. Cumulative Probability Graphs
      e. The Normal Curve and the Normal Probability Distribution

D. *Theory of Sampling; Statistical Inference*
   1. Calculations of the Distribution of a Sum
   2. The Variance of the Distribution of the Sum of Two Independent Random Variables
   3. Variance of the Sum and of the Average of Several Variables

E. *Binomial Probability Distribution and Central Limit Theorem*
   1. Binomial Experiments
   2. Expected Value of a Binomial Random Variable
   3. Binomial Probability Tables
   4. Binomial Distribution Properties
   5. The Central Limit Theorem for the Binomial

F. *Time Series Analysis*
   1. Determination of Trends
   2. Variations—Periodic, Cycle, Irregular
   3. Smoothing Techniques

G. *Control Charts*
   1. Chance vs Caused Effects
   2. Process Control
   3. Acceptance Sampling

Chapter 5
Computing Tools

I. COMPUTER GENERATIONS

The ENIAC was the first machine to use electronic tubes for calculating and was thus hailed as the first of the first generation of modern computers. Basically, vacuum tubes were used in place of the mechanical switching devices on latter-day calculators. Developed at the University of Pennsylvania by Dr. John W. Mauchly, Jr., J. Presper Eckert, and a team of associates, this machine weighed 30 tons, contained more than 18,000 vacuum tubes, and required more than 1500 square feet of floor space.

This first generation of computers also saw the introduction of rotating drum memories as the main logic element—another breakthrough. However, the machines were generally bulky, somewhat inflexible, and even required air conditioning.

The second generation began in the mid-1950s with the introduction of the transistor and magnetic core memory. These new components allowed the removal of the vacuum tubes and thus the reduction of the computer's size. Air conditioning requirements were made less strict and processing time increased. It became common to speak of processing time in terms of milliseconds (10^{-3} s) and microseconds (10^{-6} s). At the same time, high-speed printers and readers were making on-line data processing possible, and more sophisticated software and programming techniques led to widespread scientific and commercial applications.

Continued progress in electronic technology led to the introduction of the third generation of computers in 1964. This was the result of micro-miniaturized circuitry and thin-film memories. At the same time software with simplified programming procedures was bringing the computer closer to the user. Better input/output devices with random access storage opened the possibility of unlimited data storage. The impact of the third generation was reflected in new jargon such as nanosecond and picosecond (10^{-9} and 10^{-12} s, respectively).

Today, computer technology has outrun the needs, capabilities, and indeed the imaginations of its users.

The fourth generation of computers, unlike the first three, has been developing gradually with small, steady advances. The machines are modular to allow for the removal and replacement of faulty components for bench testing and repair, ensuring high reliability and trouble-free service. The modular construction also allows for expansion of a system with a minimum of cost and downtime, particularly storage expansion.

The new machines offer large, on-line data banks of information, some already loaded. Thousands of terminals are able to operate on a time-sharing basis for inquiry, data entry, updating, and output in various forms. In addition to evolutionary improvements in hardware, there are plans to develop a "fail safe" system, that is, to ensure that no individual component failure causes the system to halt.

Fourth-generation computers probably include more user orientation, short turn-around time, almost complete dependence on higher-level programming languages, faster internal speeds, larger and less expensive memories, and more complete operating systems. Also, user-oriented remote terminals, remote on-line processing, multiprocessing, and applications-oriented packages are available. Other features include the expanded use of character recognition and voice recognition. Computer manufacturers suggest that the primary characteristic of the recent generation is better price/performance ratio and more on-line communication capabilities and hardware, which assumes more software functions.

The recent machines have more logic circuits in peripherals and terminals. The systems are easier to use, have greater hardware redundancy to avoid breakdowns, and use new types of peripheral device. They have built-in subroutines, more complex instructions, low-cost but very high-speed "scratch-pad" memories, and practical associative memories. Other characteristics include greater use of logic to enhance reliability, specialized logic, various small high-speed buffer memories, wired-in spare parts and cellular redundancy, very small-size computers, and read-only memories for microprograms to imitate other computers. Microprograms within the hardware, called *firmware,* permit one computer to simulate many types of machine and thus facilitate the transition to advanced equipment.

It is interesting to note that the expectations of major computer developers are not based upon major technological breakthroughs. To be sure, the new machines will reflect advances but will also be characterized by caution and a determination to avoid mistakes previously made. Gradually systems experts, machine operators, and keypunch operators will fade from sight as the user returns to direct contact with the computer.

In view of other predictions for the 1980s, we can look forward to a generation of computer systems built to meet the economic specifications and technological limitations of the user. These machines will be aimed at the massive problems of environmental pollution, energy, and an exploding population.

II. COMMERCIAL AND BUSINESS COMPUTER LANGUAGES

In 1944, electronic computers were not commercially available. Ten years later electronic calculators were in wide use, but the electronic computer with its stored programs was still a rare item. In early 1970, we were in the midst of a generation of electronic computers characterized by solid-state, integrated circuits capable of extremely fast computing speeds. These machines are a valuable tool in the hands of the person with a problem that requires vast amounts of mathematical operations. This is true for both the scientific and business communities; both have been aided immeasurably by this device.

It is important to understand from the outset, however, the difference between scientific computer languages and those languages whose primary intent is the solution of routine business problems. A scientific computer language is used, obviously, to solve scientific problems. These involve relatively small amounts of input to the computer, small amounts of output from the computer, but large amounts of calculation by the computer during the processing stage. The main use of a scientific computer language is the reduction of data into meaningful form. Examples include matrix inversion, statistical analysis, linear programming, algebraic manipulation, numerical analysis, and so on. On the other hand, business data-processing problems are characterized by great amounts of input, great amounts of output, but minor, straightforward calculations (addition, subtraction, multiplication, division) by the computer during data processing. Business problems invariably involve files. The main purpose of the business-oriented language is to maintain files of data concerning the operation of the company. Examples include payroll processing, inventory adjustments, bookkeeping, and accounting.

A file is a collection of records. A record is a collection of detailed bits of information all related to each other in some manner. Files can be recorded on any input/output medium in use by the particular computer, or can even be maintained in a portion of the computer's memory core. Because of this heavy dependency on files, a computer language designed for business use must allow the programmer to describe in detail the records contained in the file being processed, as well as the operations to be performed. The major language presently in use that falls into this category is COBOL (COmmon Business Oriented Language). PL/1 has been touted as the one language for both business and scientific use, and it may become a replacement for COBOL in the future. It may also supplant such widely used languages as FORTRAN, ALGOL, and JOVIAL. However, the number of users of PL/1 for business applications is minimal today compared to COBOL.

The basic pattern for programming languages in business data processing (excepting PL/1) was set by FLOW-MATIC (an early business computer language), which established the concept of an English-like language with "natural" words for both the operations to be performed and the data on which they are to be done. All subsequent language developments in this area followed this concept.

A. COBOL

In May 1959, a meeting of military, business, and university leaders was called to discuss ways of developing a common business computer language, the word "common" being interpreted as meaning that the source program would be compatible with a large group of computers. The result of this and subsequent meetings was that COBOL was introduced in 1960 and has been used extensively since 1963.

COBOL is definitely not a succinct language; its objective was to be "natural" or English-like. This led to the introduction of certain concepts into the language that would permit this type of naturalness. Since one objective was to make the statements of the language easily understandable when read, some words that were logically necessary in one place became only noise words in another. The problem was resolved by defining a set of key words that were only used for specific purposes. COBOL does not permit minimal writing; on the contrary, it encourages a certain amount of verbosity. The benefit gained from this is increased readability and understanding of programs.

COBOL was designed for two subclasses of people concerned with business data-processing problems. One is the relatively inexperienced programmer for whom the naturalness of COBOL would be an asset; the other is anybody who has not initially written the program. In other words, the readability of COBOL would provide documentation to all those who might wish to examine the programs, including supervisory or management personnel. Little attempt was made to cater to the professional programmer; in fact, people whose main interest was programming tended to be unhappy with COBOL because of the large amount of writing required.

COBOL has been implemented by virtually every computer manufacturer and most of the independent software companies. One reason for this was to obtain a commercial language for customer use, but a more important reason was direct and indirect pressure from the government, which stipulated that a company that wanted to sell or rent computers to the federal government had to have a COBOL compiler unless they could clearly demonstrate that it was not needed.

COBOL has demonstrated its usefulness, and many organizations use it for all or most of their business data-processing programming. Its greatest advantages appear to be its convertability from one machine to another and its ease of use in communication and documentation. Many of its disadvantages relate to specific implementations. The early versions of COBOL were so unusable in some cases that people were reluctant to try the language. As with other languages, it continues to lack features that some people would like to have included.

Generally speaking, COBOL seems to have met most of the objectives set for it. There is no doubt that it has achieved success as a useful addition to the computing community.

B. PL/1

In 1963 a committee composed of representatives from IBM, Lockheed, Union Carbide, and Standard Oil of California was formed to remedy two inefficiencies of FORTRAN: character and alphanumeric data handling, and the inability to interact with modern equipment. The name of the new language was going to be New Programming Language, NPL. Since its initials were the same as those of the National Physical Laboratory in England, after considering MPT and MPPL, Programming Language/1 (PL/1) was finally adopted. The first manual on PL/1 was published in 1965. In August 1966, the first compiler was released for IBM System/360.

PL/1 is very general, with the widest scope of any computer language in common use today. Its notation is succinct and semiformal rather than English-like, following the pattern of FORTRAN and ALGOL rather than COBOL. It is fairly easy to read and write, with no great tendency toward errors except those due to the power and complexity of the language itself. In determining how easy it is to learn, one must specify the amount to be taught and the prior experience of the learner. A person without any experience as a programmer would find it difficult to learn all of PL/1. On the other hand, a FORTRAN programmer or a COBOL programmer would find it fairly easy to learn an equivalent amount of PL/1. PL/1 is definitely aimed at an extremely wide application area. It was meant to be used in both of the fields for which COBOL and FORTRAN were designed.

There are three major reasons for the existence of PL/1 subsets. The first is based on the stated objective of catering to or providing facilities for previous users of FORTRAN, ALGOL, or COBOL who need only comparable capabilities provided to them by PL/1. Thus, it is meaningful to talk about subsets for scientific or commercial programming. A second reason for subsets is to reduce the size of the compilers. Since this is a complex language requiring powerful compilation techniques, there will be some cases in which a smaller compiler will be desirable in handling a subsystem. The third reason, related to the second but not identical to it, involves the use of small computer configurations. The small computer with limited capacity and speed would not be able to handle all of the options allowed with the full language, whereas a specific subset of PL/1 provides for better implementation.

Within three years from the start of the PL/1 project, it had already had a profound impact upon the computing industry and caused much controversy. Its future is unknown, but there is a good chance that it will replace such major existing languages as FORTRAN, COBOL, ALGOL, and JOVIAL.

III. SCIENTIFIC COMPUTER LANGUAGES

A. Introduction

Before considering specific scientific computer languages, we should distinguish between machine languages and high-level languages. Every computer has the ability to execute a specific set of instructions. These instructions must be placed in the computer in the form of symbols that the hardware can interpret. This set of instructions must be in the direct machine language of that particular computer. Since most computers are designed so that their storage locations and registers contain binary characters, the most common machine language is a binary digit system. When using a machine language, the programmer writes the output format in very elementary, detailed steps, using the appropriate binary form of the statement in terms of zeros and ones. The programmer must also specify which registers within the machine are to be used and must keep track of the storage allocations and memory core used at each step of the program. Machine language statements are often unique to the particular computer being used; a program written for one computer may not execute on another computer.

When using a high-level language, the programmer writes the steps to be executed by the computer in terms of the computing rules of the language, not the idiosyncrasies of the computer hardware. These steps are then translated by another program, the compiler, into machine language for use by the computer. For each statement written in a high-level language, the compiler will normally create more than one instruction for the computer to execute. Thus, by using a high-level language, the programmer avoids writing the extremely large number of instructions that would be necessary with a machine language. A programmer using a high-level language does not need to know the machine code of the computer on which the problem will be run. This not only eliminates the necessity of having to think in terms of binary, an obviously unsatisfactory method of communication for people, but allows the programming on more than one computer. Thus, high-level languages are independent of the machine on which they are used, subject only to the restriction that a compiler be available to translate the instructions into machine language.

High-level languages offer other advantages when the alternative is machine language. Probably the main reason for using a high-level language is that it reduces the time and effort spent in writing and implementing the program. If a programmer chooses the proper high-level language, the notation will be very similar to the notation of the problem to be solved. This characteristic makes the problem easier to code and understand. Since the compiler takes care of the minute details of the instructions the computer will execute, the programmer has to write less in the way of instructions and can pay more attention to program logic. If there is an error in the program, it can be found more quickly since the programmer need only consider the statements required by the high-level language, not the much larger number that machine language would entail. Another advantage of high-level languages is that they provide their own documentation. When a computer program is written, comments should be inserted at appropriate places to indicate what the program is to do and what the different sections of the program are for. Since the high-level languages have a problem-oriented notation, the notation itself serves for much of this documentation. If machine language were used, the solution procedure would be expressed in an obscure, unfamiliar form, and understanding would necessarily be poor.

A final advantage of high-level languages is that they are easy to learn, and for two reasons. The programming language may be complex, particularly if one has decided to learn all facets of it, but because the notation is problem-oriented, much of the programmer's knowledge of the field can be transferred to learning the language. Secondly, since the compiler takes care of generating the internal machine language code, the programmer need only learn the grammar of the high-level language, not the internal operation of a complex electronic machine.

To be sure, there are also disadvantages to the use of high-level languages instead of machine languages. Some of these disadvantages stem from the programmer's lack of knowledge of the internal workings of the computer. This lack of knowledge may lead to an inefficiently written program or cause difficulties in correcting a program if the compiler does not provide sufficient information in its diagnostic messages. If an inefficient language is chosen to solve a problem, the programmer may not be able to express all the necessary instructions. Another disadvantage is the time required for the compiler to translate the program. Computer time is an expensive commodity. Often, the time required for compiling will exceed the time required to execute the program itself.

In the long run, the advantages of high-level languages tend to outweigh the disadvantages, although there may be exceptional cases where machine language is preferred. Opler [1] states that 98% of the scientific computation problems are solved using a high-level language.

One further distinction should be made at this point—that between professional programmers and occasional ones. Professional programmers are those whose career is programming and who are hired by firms to write computer programs to solve other people's problems. They are not the originators of the problems, but provide a service to other members of the organizations. It can be assumed that professional programmers have had the time and motivation to learn the field thoroughly, are probably intimate with the technicalities of the particular computer in use and its associated peripheral equipment, are proficient in several high-level languages as well as the computer's machine language, and should be equipped to use many of the sophisticated techniques of programming.

Occasional programmers write programs to solve problems encountered in pursuing their normal careers: engineer, systems analyst, chemist, doctor, or plant manager. They write their own programs either because there are no programmers available, or the problems are such that it would be more difficult to explain them to professional programmers than simply to write the program; it could also be that they just prefer to write the program themselves. Occasional programmers seek a language that will minimize the amount of time needed to learn those portions of it that will facilitate solution of the types of problem likely to be encountered in their jobs. Generally, they care nothing about the internal operation of the computer itself, but are concerned only with its usefulness. Our discussion will be aimed at occasional programmers.

McGee [2] has observed that "the more general the tool is, that is, the greater the number of tasks it can be applied to, the harder it is to use on any specific task." We are in the age of the general purpose computer, a computer that can be used to solve many different types of problem using many different computer languages. However, every time that we use the computer, we must have a program for it to execute. Given that we shall use a high-level language to solve our problem, which language should we use? Or to put it another way, given that we desire to use a high-level language or languages in our work, which language or languages should we learn for the type of problem that we are likely to encounter? The answers to these questions are not easy. Many factors must be taken into consideration. In the following sections we shall discuss two types of language of interest to the scientific community: procedure-oriented languages and application-oriented languages.

B. Procedure-Oriented Languages

When using a procedure-oriented language, the programmer specifies a sequential set of executable operations that the computer is to perform. The solution is written in terms of the *algorithms* (computational procedures) that the computer executes. The key factor here is that the statements written by the programmer are definitely executable operations and that the sequence of execution of these operations must be specified. For example, the following might be part of a program written in FORTRAN:

$$A = 1.0$$

$$B = 3.0$$

$$C = A + B$$

The computer would first set A equal to 1. It would next set B equal to 3. Finally (in this sequence), it would add A and B to get 4 and then set C equal to this value. In general, then, we can see that a procedure-oriented language requires the user to indicate to the computer the *exact* operations that it is to perform on the data or information it has been given, and in what sequence.

1. Languages for Numerical Scientific Problems

Languages designed for the computer solution of scientific problems using numerical techniques are the best-known scientific programming languages. These use algebraic expressions to instruct the computer in the solution of the problem and are characterized by relatively small amounts of input to the computer (usually the program itself and the numerical data needed to solve the problem), large amounts of calculation during the processing, and a small amount of output. The intermediate answers are usually of little or no interest to the programmer, so the computer's output is considered to be small in comparison with the amount of work needed to obtain the solution. The following is a review of some of the languages that fall into this category.

a. FORTRAN

The development of FORTRAN and the beginning of the development of computer programming using high-level languages were almost parallel. In describing the history of this language, Sammet [3] states:

> The earliest significant document that seems to exist is one marked "PRELIMINARY REPORT, Specifications for the IBM Mathematical FORmula TRANslating System, FORTRAN," dated November 10, 1954, and issued by the Programming Research Group, Applied Science Division, of IBM. The first sentence of this report states, "The IBM

Mathematical Formula Translating System, or briefly, FORTRAN, will comprise a large set of programs to enable the IBM 704 to accept a concise formulation of a problem in terms of a mathematical notation and to produce automatically a high-speed 704 program for the solution of the problem." It is interesting to note that the authors (who are not identified in the document) felt a need to justify such a development. They devoted several pages to a discussion of the advantages of such a system. They cited primarily the virtual elimination of coding and debugging, reduction in elapsed time, doubling of machine output, and the feasibility of investigating mathematical models.

The system mentioned above refers to the development of a compiler for translating FORTRAN into the machine language of the IBM 704 computer. The 704 FORTRAN system was introduced in early 1957.

Although FORTRAN and other high-level languages are now quite common, the 704 system was not readily accepted in its early days. Many objections to this system were raised, mostly based on the contention that the compiler could not turn out a machine language program as well as the best programmers. It took a significant selling campaign to get FORTRAN implemented on a wide scale.

Another problem also occurred as a result of the introduction of the FORTRAN compiler. Sammet [3] reports:

In some sense, its introduction caused a partial revolution in the way in which computer installations were run because it became not only possible but quite practical to have engineers, scientists, and other people actually programming their own problems without the intermediary of a professional programmer. Thus the conflict of the open versus closed shop became a very heated one, often centering around the use of FORTRAN as the key illustration for both sides. This should not be interpreted as saying that all people with scientific numerical problems to solve immediately sat down to learn FORTRAN; this is clearly not true but such a significant number of them did that it has had a major impact on the entire computer industry.

FORTRAN is considered to be rather easy to learn. It is rare that a course in this language runs for more than 1 or 2 weeks, and many technically oriented persons have learned enough in 1 day to continue on their own. In addition, many people have learned the language without the benefit of formal classroom instruction. Programmed instruction and correspondence courses are available for those with the interest and time to pursue FORTRAN by themselves. As an example of the ease with which FORTRAN may be learned, and the advantages of the language over machine language, Rosen [4] reports:

The experience of the FORTRAN group in using the system has confirmed the original expectations concerning reduction of the task of problem preparation and the efficiency of output programs. A brief case history of one job done with a system seldom gives good measure of its usefulness, particularly when the selection is made by the authors of the system. Nevertheless, here are the facts about a rather simple but sizeable job. The programmer attended a one-day course on FORTRAN and spent some more time referring to the manual. He then programmed the job in four hours, using 47 FORTRAN statements. These were compiled by the 704 in six minutes, producing about 1,000 instructions. He ran the program and found the output incorrect. He studied the output (no tracing or memory dumps were used) and was able to localize his error in a FORTRAN statement he had written. He rewrote the offending statement, recompiled, and found that the resulting program was correct. He estimated that it might have taken three days to code this job by hand, plus an unknown time to debug it, and that no appreciable increase in speed of execution would have been achieved thereby.

This case occurred about 1957. Since then many improvements have been made not only in the language of FORTRAN itself, but in the speed with which these programs can be compiled for the computer. Nevertheless, at the time the reduction in programming time from three days to four hours, and the reduction in the number of required statements from about 1000 in machine language to 47 in FORTRAN was quite significant. As this programmer gained experience, he probably shortened the programming time even further and learned more sophisticated techniques allowing him to reduce the number of statements he had to write.

A FORTRAN compiler is available on almost every computer in use today. It is widely accepted as a general purpose mathematical and engineering language that can be applied to a large spectrum of problems, including some limited business data processing applications. Its statements are similar to ordinary mathematical notation, which partly accounts for its great usage by the scientific community. It should not be inferred that FORTRAN is limited in use to the mathematician or engineer. The language is applicable to many disciplines where the solution to a problem involves considerable mathematical manipulation.

b. ALGOL

The development of ALGOL (ALGOrithmic Language) was unique in that it was the first major computer language to be designed by a committee of representatives from several international organizations. Work on ALGOL [at first known as IAL (International Algebraic Language)] began in 1955 when the German Association for Applied Mathematics and Mechanics formed a committee on computer programming. This committee set up a subcommittee to investigate formula translation and the construction of a compiler. In May 1957, a conference was held in Los Angeles, California, to examine methods of facilitating the exchange of computing information. The conference resolved that the development of a single universal computer language would be very desirable. By October, because of the existence of many other computer languages, it was decided that an effort should be made to unify these languages. An ad hoc committee composed of representatives from computer users, computer manufacturers, and universities was formed. In January 1958, this committee discussed the technical details of a programming language. As a result of these and other meetings, the following objectives were agreed upon [3].

1. The new language should be as close as possible to standard mathematical notation and should be readable with little further explanation
2. It should be possible to use it for the description of computer processing in publications
3. The new language should be mechanically translatable into machine programs

It was implied in the committee reports that the language was to provide a standard. The results of all these efforts was ALGOL 58, a publication language designed for the dissemination of computing subroutines. For this reason, there were no facilities provided in the language for input or output from a computer.

At the UNESCO-sponsored International Conference on Information Processing held in Paris in June 1959, an open discussion of the weaknesses of ALGOL 58 was held. It was agreed that another meeting would be held in Paris in January 1960 to discuss improving the language. This meeting was attended by 13 representatives from Denmark, England, France, Germany, Holland, Switzerland, and the United States who worked for agreement on every item discussed. The resulting report delineated the basis for ALGOL 60, which provided input/output facilities.

ALGOL was intended to be used in the same types of application as FORTRAN, but greater emphasis was placed on the sophisticated mathematical procedures. ALGOL was designed to express algorithms for the solution of a wide variety of problems, although it has the disadvantage of not being as easy to learn as FORTRAN.

When first presented to the world, ALGOL was proclaimed to be the wave of the future. However, its application has been limited mostly to European users and universities, although several U.S. computer manufacturers have developed compilers for this language. Perhaps one reason for the lack of support for ALGOL in this country is the failure of IBM to back the language. IBM has been the dominant computer manufacturer and as such can to some extent determine the success or failure of any innovations in the field. Although IBM now supplies an ALGOL compiler, in the early 1960s the firm was enthusiastically promoting its own successful development, FORTRAN.

c. PL/1

It was realized early on that although FORTRAN was efficient for solving scientific numerical problems, it was not effective for handling alphabetic and alphanumeric data. The many improvements that were incorporated in FORTRAN did not solve those particular problems. In September 1963, a committee was formed to develop a language useful to a wide range of people and at the same time an effective tool for the engineer. Originally the purpose of the committee was to improve FORTRAN by merely extending it in some areas to remove some of the restrictions that it placed on the programmer. The committee soon realized that compatibility could not be retained with FORTRAN while developing a language that could meet modern programming requirements. It was finally decided that the new language would *not* be compatible with FORTRAN.

The committee considered various features that might be included in the new language. The committee also studied other languages to find capabilities that might be included in the new language, even if in a different form. All reasonable attempts were made to consider ideas from as wide a scope as possible. The committee's findings were released in March 1964.

At first glance, the new language, PL/1 (see also Section II.B), appears to be just an extension of FORTRAN. However PL/1 is much more powerful. It was the first language to consider the problems of the programmer with respect to the computer. The PL/1 compiler makes every reasonable attempt to run a program, making assumptions where needed in order to execute it. The advantage of this is that the program is not rejected on a minor technicality. At the same time, however, a programmer who is not thoroughly familiar with the language may find a program being executed differently from the way it had been planned because of the "default" conditions. This is not a significant concern, however, because the compiler will in most cases provide a diagnostic message indicating the assumptions that it has made in the execution of the program. PL/1 also incorporates many capabilities and features that are not found in FORTRAN, giving it a wider potential to solve scientific problems. It is not our purpose to discuss these advantages here; it is sufficient to say that PL/1 allows the user greater flexibility in solving highly complex notational problems than does either FORTRAN or ALGOL.

Rubey [5, 6] compared PL/1 with other programming languages by devising several programming problems in the areas of business, science, simulation and gaming, human–machine interaction, and data management for professional programmers with from one to ten years of experience. They coded each problem twice, once in PL/1 and once in another language. Rubey made the following observations [6]:

> In almost every case the programmers who compared PL/1 and FORTRAN
> responded that PL/1 was superior. In expressing their preference for PL/1,

they cited the fact that the capabilities of PL/1 permitted direct solution
of the problem, whereas little tricks or clever manipulations were required to accomplish the same tasks in FORTRAN. It was concluded that
there is little to choose between PL/1 and FORTRAN for a small scientific
problem consisting mainly of equations. For larger scientific problems
and for problems in other application areas such as the development
of interactive programs or simulations, PL/1 was considered superior
because of its input flexibility, logic control features, and character and
bit processing capabilities. The free form of its statements and freedom
in symbolic label selection make PL/1 more self-documenting than FOR-
TRAN.

Although the professional programmers felt that PL/1 was superior for
their own needs, Rubey concluded that "because of its simplicity, FOR-
TRAN is easier to learn than PL/1 and hence more suitable for the non-
professional programmer" [6].

PL/1 has been advertised as the one computer language for both the
scientific and the business communities because it has the ability to handle
not only the complex mathematical operations of the scientist, but also
the character strings and texts of business applications. The design of PL/1
was based on the concept that one language could serve the needs of both
areas of data processing. The significance of the language's development,
thus, is that it is a synthesis of the best of many other languages, with
most of their advantages and few of their disadvantages. Because PL/1
was created after years of experience with such languages as FORTRAN
and ALGOL, the designers were able to avoid many of the problems that
had been encountered with these languages.

d. BASIC

The BASIC (Beginner's All-purpose Symbolic Instruction Code) language
was developed at Dartmouth College in 1965 under a grant from the National Science Foundation. The project was directed by Professors John
G. Kemeny and Thomas E. Kurtz. BASIC was designed to be a simple
language to learn and an easy language to translate. In addition, its developers hoped it would become a stepping stone for students who were
attempting to learn one of the more sophisticated languages such as FOR-
TRAN or ALGOL. BASIC has been implemented on most computers manufactured by the General Electric Company.

Computer programs can be submitted to the computer in essentially
one of two ways. The oldest method is the "batch" type of processing
in which the programmer punches the program on cards, delivers the data
cards to the computer facility, and returns at some later time to receive
the output. The newer method, and one that is gaining in popularity, is
the on-line procedure. Under this method, the user sits at a typewriter-
like console and uses the computer directly, even though many others
may be using it at the same time. The advantage of the on-line procedure

is that the programmer gets an answer almost immediately and does not have to wait until the computer center has an opportunity to run the program. If there are errors in the program, the user can correct them immediately.

BASIC has been accepted by a significant segment of on-line computer users. Szuprowicz [7] reports the following results of a survey on computer languages and the percentages of people who use them for on-line applications:

FORTRAN	39.6%	COBOL	7.6%	ALGOL	2.8%
BASIC	31.0%	PL/1	7.0%		

The remaining 12% used another of the large number of on-line languages available, none of which accounted for a significant percentage of users. COBOL, mentioned in the tabulation, is a business language with very little usefulness to the scientific computing community (see Section II.A).

As stated earlier, BASIC was designed to be easy to learn. The language is well suited to computer-aided instruction at the same console at which it is used. Student errors are detected immediately, and test programs can be run simply and quickly. Sharpe [8] points out that "the average college student should be able to learn it with five to ten hours of class-room exposure and an equal investment in time spent preparing actual programs." In many cases, the language can be learned by studying a text, thereby eliminating the actual classroom time.

e. APL

The computer language APL (A Programming Language) had its beginning in 1957 when Dr. Kenneth E. Iverson began work on a book on computer applications. Dr. Iverson's original intent was not to create a computer language, but to develop a mathematical notation that would clarify his work. On reviewing the new system of notation, he found that it could be applied to many fields of computer application.

APL is an algebraic language with unusual, but powerful, operational symbols. To use the language, programmers must ignore some of the mathematical conventions they have learned. For example, the order in which operations are executed is from right to left in all cases and expressions. Another APL convention is that all operators are placed before their operands, whereas in ordinary algebra the operators may be before the operand (like the plus or minus sign) or after the operand (like the factorial sign). This is not to imply that the operators of the language are the common operators known by every high school algebra student. APL incorporates many unusual operators, such as arrows (that can point up, down, to the left, or to the right), Greek letters, triangles, rectangles, and a host of others. Each symbol has a compact meaning, the use of which saves many statements in other language. Once learned, APL notation is highly consistent and may be used to represent complex relationships and overall logic concisely and unambiguously.

APL has been implemented in an on-line interpretive system by IBM. The term "interpretive" means that the system does not wait until a program has been completely fed into the computer before the compiling stage takes place; each statement is interpreted as it is written on the console, compiled into machine language, and immediately executed. This system reduces the time the programmer spends waiting for output.

The computer world is presently divided on the subject of APL. There are those who think it is the answer to all problems and those who see no use for it at all. In the middle are those who see many of the advantages of the language but still recognize its limitations. The outstanding characteristic of APL is its heavy use of operators on vectors and matrices. However, many people cannot become enthusiastic about a language whose notation is so complex. The proponents of APL feel that once an understanding of the notational system is developed, the language will be easy to learn. There is no doubt that it is a powerful language and more efficient per statement than any other computer language in use today. However, for the occasional programmer, APL would probably require more instruction time than it is worth.

2. Languages for Formal Algebraic Manipulation

Languages designed for formal algebraic manipulation are used for the computer processing of formal mathematical expressions without concern for their actual numerical values. The types of operation involved include differentiation, integration, and substitution of algebraic expressions.

It may not be immediately obvious why this type of work requires the use of a computer. Consider that in the solution of many mathematical applications there is a tremendous amount of extremely tedious mechanical algebra. This algebra is as susceptible to human error as are numerical calculations. There are cases in which months worth of mathematical manipulation "by hand" were found to be wrong when checked by a computer. Just as the computer can save time and trouble in the solving of numerical problems, it can also be used as an aid in the solution of symbolic algebra and calculus problems. The following sections discuss two languages that are used in this type of computer work. Although these languages are considered to be special purpose or application oriented by many writers, they are included here because their concepts are important and because they can be used to perform a wider range of manipulations than those classified as strictly application oriented (see Section III.C).

a. FORMAC

FORMAC (FORmula MAnipulation Compiler), an extension of FORTRAN, is designed for the manipulation of mathematical expressions. Its basic concepts were developed in July 1962 by Jean E. Sammet and Robert G. Tobey, working at IBM's Boston-based Advanced Programming Department. At the time it was recognized that there was a need for a formal algebraic capability associated with a language already in existence, and FORTRAN was the logical choice. The first draft of the specifications for FORMAC was completed in December 1962, and the design implementation was begun shortly afterward. The basic goal of this work was to develop a formal mathematical manipulation system for the IBM 7090/94. Originally FORMAC was intended to be merely experimental and available only within the IBM Corporation. By April 1964, after thorough testing, the first complete version of FORMAC was in operation. In November 1964, FORMAC was released to a number of IBM customers. Even though it was considered to be experimental and lacked the "official" status of other compilers, FORMAC was used extensively. At about the same time, work was begun on associating FORMAC with PL/1, with the aim of implementing it on the IBM System/360. The PL/1–FORMAC system was released by IBM as a contributed library program in November 1967. The system incorporated many of the concepts of the original FORMAC, but also included many improvements and additions.

In the opinion of the developer, the most significant contribution of FORMAC is that it introduced the concept of adding a formal mathematical manipulation capability for solving numerical scientific problems to a language already in existence [3]. FORMAC also proved that a practical system could be developed to do formal algebraic manipulation on a computer, and that the language could be easily learned and used to solve specific engineering and mathematics problems. There was one major drawback discovered during experimentation, however: Lack of computer internal storage could limit the problems that could be solved. Many algebraic expressions require a large amount of space in the computer. As the expressions grow in size, the storage requirement grows by an exponential function; thus, doubling the space available does not double the effectiveness of the computer to handle such a problem. Still, even with their present weaknesses, FORMAC and PL/1–FORMAC are convenient tools for the person with a problem involving the manipulation of mathematical functions.

b. MATHLAB

C. Engleman began developing MATHLAB, an on-line system for formal algebraic manipulation problems, in 1964. MATHLAB has since undergone further development. The original version was replaced by MATHLAB 68, a system that became operational in the fall of 1967.

MATHLAB was the first language to include high-level operations such as integration, and its most significant contribution is considered to be the development of routines that could complete these operations. MATHLAB is a valuable tool to the scientist with access to an on-line console that incorporates the system. Consider that within seconds the solution to a complex set of mathematical expressions can be obtained merely by sitting at the typewriter and feeding the information into the computer. Equations that could take days to solve can thus be reduced to meaningful form with minimal human effort.

C. Application-Oriented Languages

As used here, the term *application-oriented* applies to those languages whose primary purpose is the solution of particular types of problem. Although the distinction between this class of languages and the procedure-oriented languages previously discussed is not absolutely clear-cut, application-oriented languages are generally more limited in scope and provide the programmer with less control over the actions of the computer. Most of these languages have built-in subroutines that perform specific tasks. The control exerted by the programmer in designing the solution is strictly one of determining which subroutine to employ. Thus, when using an application-oriented language, the programmer describes a problem in terms of the language itself. The solution procedure is then predetermined by the language and the associated subroutines. Any particular application-oriented language is useful only for a certain family of problems.

Application-oriented languages are generally more powerful than the procedure-oriented languages when the problem under consideration is one for which the application-oriented language has been designed. For example, it is quite possible to write a program in FORTRAN to simulate a manufacturing job shop. However, by using one of the simulation languages (see Section III.C.1), the time and effort expended in the solution to the problem is considerably reduced. A linear programming problem is almost trivial when the programmer uses IBM's Mathematical Programming System, but could become complicated when the same solution is attempted in PL/1.

With this distinction in mind, then, we shall investigate three categories of application-oriented language: simulation languages, languages for certain common applications, and specialized technical application languages.

1. Simulation Languages

Simulation implies the *dynamic* modeling of some kind of system. The system may be characterized by discrete or continuous events. Computer simulation provides an effective and efficient method of testing and evaluating a proposed system under various conditions without

1. Constructing a physical and/or environmental model, if they are proposed systems
2. Disturbing them if they are operating systems with which it is costly to experiment otherwise
3. Destroying them, if determination of their maximum stress is the purpose of the experiment

Several days, weeks, or even years of simulated events can be performed on a computer in a matter of minutes. The results obtained can be invaluable in gaining new insights and establishing the feasibility of new ideas and alternatives.

Computer simulation is not an exact analog of the real system. What is obtained is the behavior of a model of the real system. This requires the analyst to exercise careful judgment both in the construction of the model and in the interpretation of the simulation results. Computer simulation may also allow measurements that would be physically or economically impossible to obtain any other way. This ability immeasurably enhances the value of computer simulation as a tool for the business manager, engineer, or functional specialist. It also accounts for the fact that simulation is being used for an ever-increasing number of applications with very useful results.

The common characteristic of all simulation endeavors is the construction of a model. The use of models has for many years been an important technique in the solution of engineering and scientific problems. The first step in carrying out computer simulation is to construct a model that represents the system being studied. The degree of validity of a simulation model is highly related to the degree to which the model duplicates the real system. However, the closer the model to the real system, the more complex and expensive the simulation process. The decision regarding this trade-off should be made with respect to the purpose of simulation study and the criticality of the accuracy of the simulation results. Statistical analysis is required both in data acquisition and sampling and in the analysis of the results.

Simulation models may be classified into two major categories: discrete change models and continuous change models. A *discrete change model* is appropriate when the system contains elements that perform definite functions. Items flow through the system from one element to another, and these elements have a finite capacity for processing the items. Because of this finite capacity the items may have to wait in *queues*. The usual goal of studying such systems is to determine the capacity of the system to handle the items. The results of a series of this type of simulation will indicate how many of the items can pass through the system in a given period as a function of the parameters of the system. The statistics compiled as a result of the simulation are usually calculated using queueing and probability theories and represent the behavior of the simulated system during the run.

A *continuous change model* is appropriate when the programmer conceives of the system being studied as consisting of a continuous flow of information or material rather than as a flow of discrete items. These models are simulated with the use of finite-difference equations which approach the differential equations needed to represent a continuous flow. Chemical process, water resource, and population and economic growth systems are some examples of systems for which continuous simulation is applied.

A simulation language can be characterized by the type of world view that it employs. Generally, discrete simulation languages employ either process- and/or event-oriented world views. In a *process-oriented simulation language,* a sequence of events and activities that are commonly observed in many systems is defined by a statement. The language utilizes a set of these statements to model the behavior of the system. Such simulation languages are generally easy to use and require no prior knowledge of programming, their basic disadvantage being a lack of flexibility to model complicated and uncommon processes. An *event-oriented language* simulates the behavior of the system by scheduling the events that change the state of the system and that can take place at isolated points in time. In such languages, the simulation time is advanced from one event-occurrence time to the next while updating the state of the system as the result of such occurrences. The simulation modeling with these languages is usually more involved and may require a prior knowledge of a procedure-oriented language. Their main advantage is their greater flexibility in modeling.

A brief discussion of the process-oriented, event-oriented, and continuous simulation languages follows. None of the simulation languages discussed is best for all purposes. Usually the programmer must use the language available on a particular computer. However, these languages are quite powerful, and even though the language that is used might not be optimal for a given application, a little ingenuity on the part of the analyst will very often lead to excellent results.

a. GPSS

The General Purpose Simulation System (GPSS) was one of the earliest simulation languages developed and was first introduced to the public in 1961. It is an IBM computer program for conducting simulated evaluations of any type of discrete system. In contrast to most simulation languages, GPSS programs are based on a *block diagram* approach. In brief, the user draws a flow chart of the system to be represented. Each block in the flow chart represents a process in the system, and changes in the system are caused by a series of transactions that flow through the blocks at specified time intervals. The connections between the blocks indicate the primary and alternative sequences of processes, with the conditions of the system at the time determining which path is chosen.

The system being simulated will usually consist of items of equipment with which the transactions interact as the customer moves from one block to another. There are two basic types of equipment available. The first is called a *facility*, and can only accommodate one transaction at a time. The second type is *storage* and can accommodate many transactions simultaneously, the limit on the number of transactions being set by the programmer. [To clarify the point, one may consider a single server before whom customers line up to receive service. While the server (facility) can handle only one customer at a time, the waiting area (storage) is capable of handling many customers simultaneously.] In addition, certain attributes of the transactions or the system itself can be specified and measured by the program at different time intervals. These attributes are used to make certain decisions in the course of the execution of the program. Output statistics from the simulation run may be presented in a form specified by the user and may be obtained in a graphical representation if so desired. There are many "automatic" features of GPSS that furnish these statistics in a predetermined format, so there is relatively little need to be concerned with the input/output statements so common to other types of language. The automatic output from a GPSS program provides information on the following [9]:

1. The amount of transaction traffic flowing through the complete system and/or any of its parts
2. The average time it takes for transactions to pass through the complete system or between selected points, and the distribution probability of this passage time
3. The degree to which each item of equipment in the system is loaded, together with its distribution
4. The maximum and average lengths of queues occurring at various points, as well as their distribution

GPSS is relatively easy to learn. Its simplicity derives from the fact that the creation of the transactions, events, and activities, as well as the output, are almost completely programmed for the user. In addition, the criteria for the sequence of events and the conditions of the system are imbedded in the block diagram itself, and little extra programming effort is required. GPSS restrictions derive from the loss of flexibility that comes from making the language as easy to use as it is. Certain activities and conditions are difficult, if not impossible, to describe because the system characteristics have been limited to a set of fixed concepts. Moreover,

GPSS is a discrete simulation language and therefore not readily adaptable to many problems requiring continuous functions.

Even with its limitations, GPSS can be applied in most industries to solve a wide variety of problems. Generally, these problems have one common characteristic: competition among people, or equipment for the services of other people or equipment. The problem is to determine how well the system will respond to the demands anticipated. GPSS can offer direct assistance in analyzing this type of problem.

b. SIMULA

SIMULA is a process-oriented discrete simulation language that is used more widely in Europe than in the United States. Like SIMSCRIPT (see Section III.C.1.d), it was conceived as an extension of a general purpose language, in this case ALGOL. In SIMULA, the system is viewed as sets of processes, and simulation is accomplished by program blocks that effect these processes. It is felt that this language, particularly in its implementation as SIMULA 67, is an elegant and powerful discrete simulation language. However, because there is little interest in or availability of ALGOL and ALGOL-based languages in the United States, the use of SIMULA almost certainly will continue to be inhibited.

c. Q-GERT

Q-GERT, originally developed by Pritsker [10], is a process-oriented simulation language that introduces a network representation (rather than a block representation as in the case of GPSS) for ease of model building. GERT is an acronym for Graphical Evaluation and Review Technique; the prefix Q indicates that queueing systems can be modeled in graphic form. A fundamental contribution of Q-GERT is its method for graphically modeling systems in a manner that permits direct computer analysis (with the Q-GERT Analysis Program).

Q-GERT employs an *activity-on-branch* network philosophy: A branch represents an activity that models a time-consuming process and nodes that model events (milestones), decision points, and processes. Flowing through a Q-GERT network are entities referred to as *transactions*. A simulation model in Q-GERT is therefore represented by a network that includes a number of nodes with leading and emanating branches to and from the nodes. As transactions move along the branches, the simulation time is advanced by the travelling time of the entity, which is specified on the branch. As the transaction reaches a node it indicates the occurrence of an event that causes a change in the state of the system. Since different events may have different effects on the state of the system, a number of nodes with different characteristics, each representing a particular event, has been specified in Q-GERT.

Recent developments in Q-GERT involve the use of a graphic terminal for constructing Q-GERT networks and displaying them on a terminal screen. A displayed network can be edited, modified, and sent as an input automatically directed to the Q-GERT Analysis Program. This capability can also serve as a visual aid in tracing the movement of the transactions through the network as the Q-GERT analysis program is being executed. This latter real-time approach may be found as a good teaching tool in simulation courses.

d. SIMSCRIPT

SIMSCRIPT was developed in the early 1960s by the Rand Corporation. Like GPSS, SIMSCRIPT is designed for the simulation of discrete systems. Its underlying concept is that a system can be described in terms of certain entities, the properties of these entities, and the grouped entities themselves. The language resembles FORTRAN in appearance, probably because it was first used to translate FORTRAN programs for simulation.

While GPSS is oriented to the flow or block diagram approach, a SIMSCRIPT program is based on statements similar to programs written in the procedure-oriented languages already discussed. This feature gives SIMSCRIPT a more intensive capability for the construction of models than is obtained with GPSS. This means it can be used to represent significantly more complex systems but in turn requires a greater level of programming skill. SIMSCRIPT is considered to be one of the most powerful simulation packages available and is capable of utilizing almost the full capacity of a computer in the analysis of a simulation model.

Recently, new modeling concepts have been incorporated into SIMSCRIPT II.5, which enable the user to combine the flow of the block diagram approach (process orientation) of GPSS and the event-oriented approach of the earlier version. SIMSCRIPT II.5 has several advantages:

1. It provides a much more natural framework for modeling of complex systems
2. It is much more teachable
3. The amount of programming is significantly reduced

e. GASP

GASP was originally developed at U.S. Steel in 1963 by Philip J. Kiviat. A second version, GASP II, was developed at Arizona State University during 1964–1969. The most recent development has been GASP IV, which was developed by Alan B. Pritsker of Purdue University. GASP represents a concept in simulation languages completely different from that offered by GPSS and SIMSCRIPT because it is written in FORTRAN and can therefore be recompiled using any FORTRAN compiling system available.

GASP utilizes an event-oriented modeling world view. In this orientation the system is modeled by defining the changes that occur at event times. The primary task of the models is to determine the events that can change the state of the system and then develop the logic associated with each event type. The user-written subroutines incorporate such logics. GASP analysis program provides a filing system that allows the execution of the logic associated with each event type in a time-ordered sequence. Capabilities provided by GASP are as follows: generation of random variables to determine event occurrence times; filing, removing, and searching for events according to any desired logic; and gathering and summarizing information that describes the performance of the system and its variables. GASP provides very powerful data-gathering potential and, because it is FORTRAN based, lends itself quite easily to most computer systems. GASP IV can also provide joint discrete/continuous capabilities.

f. DYNAMO

DYNAMO, one of the earliest simulation languages, was developed at MIT and first implemented in 1959. This language is used for continuous systems, meaning that every basic variable within the system is continuous and has a first derivative with respect to time. Like SIMSCRIPT, DYNAMO is statement oriented and requires more programming training and experience for effective use than does GPSS.

DYNAMO is applicable to a continuous system when each variable and all its derivatives with respect to time exist and are known at any moment in time. This language also incorporates auxiliary equations for the inclusion of complicated variables and arbitrary functions. The results of a DYNAMO program are obtained by deriving a sequential solution of all the equations describing the simulated system. The major use of this language so far has been in the area of research. Unlike SIMSCRIPT, however, the ease of its use does not depend on prior knowledge of a procedure-oriented language.

g. CSMP

The Continuous System Modeling Program (CSMP) was developed by IBM to satisfy the need for a problem-oriented program that could be used on a large-scale computer without prior knowledge of computer languages. CSMP is a program for the simulation of continuous systems that accepts problems expressed either in the form of analogue block diagrams or a system of ordinary differential equations. Within the language is a set of basic functional blocks by which the continuous system may be represented. Application-oriented input statements are provided to describe the interaction among these functional blocks. If the programmer knows FORTRAN, the versatility of CSMP is enhanced. By using FORTRAN statements, the program can be adapted to even more complex problems, since this allows the user to deviate from some of the structured parts of the CSMP language itself.

CSMP was specifically designed with the scientist or engineer in mind.

It requires a minimum of knowledge of computer programming and operation, is simple yet flexible, and allows the user to concentrate on the system being studied rather than the sometimes formidable mechanics of computer programming.

h. SLAM

The Simulation Language of Alternative Modeling (SLAM) is one of the most recent and most advanced simulation languages developed by Pritsker and Pegden [11]. This FORTRAN-based language combines the process-oriented, event-oriented, and continuous-modeling world views into a single framework, while allowing independent utilization of each of the three modeling alternatives. Therefore, while having the advantages of each alternative, it is potentially free of their disadvantages. SLAM provides network symbols for building graphical models that are easily translated into input statements for direct computer processing.

The process-oriented part of SLAM is very similar to the Q-GERT language. The network orientation of SLAM, however, includes a number of additional node types which allow an increased flexibility for modeling of more complex systems.

The discrete event-oriented and continuous-modeling aspects of SLAM are essentially taken from GASP IV, with minor modifications. The strength of SLAM basically lies in its ability to interrelate these three modeling aspects and to be used efficiently to simulate any possible system with any degree of complexity.

One version of SLAM includes an optimization routine that interfaces with the SLAM main processor and provides a strong tool for the analysis of system performance for a large number of system specifications alternatives generally practiced for design purposes.

Future versions of SLAM will possibly include such revolutionary features as a data base system and a built-in statistical analysis package to support simulation modeling and analysis activities.

i. COMPARISON OF SIMULATION LANGUAGES

The decision of which computer language to use for writing a simulation program arises in two contexts:

when an analysis group which is likely to use simulation techniques is formed

when an individual analyst is about to prepare a simulator

Any formal computer language has many characteristics. The relative importance of these characteristics depends on whose interests are being considered. For example, the learner's interests differ from the user's, which are different from those of the system operator. Although we shall try to include the considerations relevant to each of these groups, our primary consideration is the user's point of view. The user wants a language that

facilitates model formulation

is easy to program

provides good error diagnostics

is applicable to a wide range of problems

The first criterion requires that a language be problem oriented; that is, the commands and data designators in the language should be the same as those used in the analyst's native language. The second and third criteria are partly a function of the problem orientation of the language and partly of the cleverness with which the translator is constructed. The last requires that any sort of state change that we might desire to represent can be represented in the language.

Let us look at the eight simulation languages previously described (as well as JOB SHOP SIMULATOR, which represents a variety of job shops by means of parameters) in terms of problem orientation, error detection, and general applicability. These languages can be classified in terms of orientation and scope or generality of application as shown in Figure 1.

The trade-off between generality (breadth of application) and problem orientation is clear. FORTRAN is included in the figure as an example of a multipurpose language in which any sort of state-change process can be described. Even though it is not a simulation language, FORTRAN can

FIGURE 1. Classification of simulation languages (relative location).

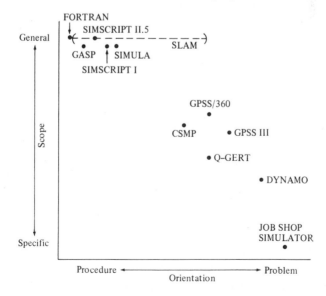

be used to write simulation programs. ALGOL and PL/1 are also suitable in this sense; PL/1 has list-processing features that make it particularly suitable for simulations.

GASP and SIMSCRIPT differ in that the former is not a complete language. Both languages can do anything that can be done in FORTRAN. Although they offer valuable basic assistance in preparing simulators, they still demand that the analyst make most of the translations from the problem formulation to the state-change model as well as be concerned with some programming details. To illustrate how simulation languages improve, an early version of SIMSCRIPT, SIMSCRIPT I, is shown in the figure.

GPSS/360 is very much oriented toward a particular kind of problem; however, it has many features that permit it to be applied in a wide range of situations. Furthermore, the language can be augmented by subroutines written in Assembly Language. GPSS II, an early version of GPSS, is also positioned in the figure to show the language's improvement.

DYNAMO was developed for defining models of business and CSMP for engineering design applications; however, their applications could be interchanged. Neither language is very general, but both are quite useful in specifying simulation procedures for specific types of problem.

JOB SHOP SIMULATOR is so specific that it is not a language at all, but a program for simulation of a particular problem. It is general only to the extent to which the size of the problem (e.g., number of machines) and some details of the operations (job routine rules, dispatching decisions) can be varied. Such packaged programs are not used extensively in simulation, since languages are much more general and are almost as easy to use for a particular problem as is a special program.

A broader comparison of the seven simulation languages (excepting SIMULA, which has little application in the United States) is summarized in Table 1.

2. Languages for Certain Common Applications

Often a problem arises whose solution requires a common mathematical or analytical technique. The source of the problem is not important to our discussion, since the procedure used to solve it could be applied to problems from a wide range of disciplines. Matrix algebra or linear programming are two techniques that can be used in a number of different areas. We shall now consider a sampling of the types of language that are helpful in the solution of a particular class of problem and do not need a program written in one of the procedure-oriented languages.

a. MPS

IBM's Mathematical Programming System (MPS) is designed to solve systems of linear or separable functions using the accepted techniques of

TABLE 1. Comparison of Seven Simulation Languages[a]

Language	Flexibility and power for system-state modification					Language use				
	Relative degree of applicability	Data storage/retrieval flexibility	Arithmetic operation capability	Set operation capability	Logical test and program continuity	Prior language requirements	Degree of ease	Output flexibility and utility	Speed of execution[b]	Operational memory size
GPSS	M[c]	L	M	L	Yes	None	H	L	H[d]	Large
GASP	H	M	L	M	Yes	FORTRAN	L	M	H	Small
DYNAMO	L	VL	VL[e]	VL[f]	No	None	H	L	H	Small
SIMSCRIPT	H	H	L	H	Yes	FORTRAN	VL	M	H	Large
CSMP	L	VL	VL[e]	VL[f]	No	None	H	L	H	Small
Q-GERT	M	L	M	M	Yes	None	H	L	H	Small
SLAM	H	H	H	H	Yes	None or FORTRAN	L	M	H	Small

[a] Derived from author's personal experience and from Krasnow and Merikallio [12].

[b] Although speed of execution can best be evaluated by extensive testing of a standard model on different simulators, no such comparison is available.

[c] Code: H, high; M, medium; L, low; VL, very low.

[d] According to Mize and Cox [13], GPSS is somewhat slower than SIMSCRIPT.

[e] Only on fixed equation.

[f] No set-forming capability.

linear or separable programming. *Linear programming* is a technique used to solve a set of linear inequalities. *Separable programming* is used in solving certain types of nonlinear function within the framework of the general linear programming procedure. Problems of these types often occur in such areas as machine loading, materials allocation, ingredient blending, distribution and shipping scheduling, labor allocation, and so on. To use MPS, the analyst first builds a model of the problem in terms of the equalities or inequalities that express the objective of the system and the constraints that have been imposed on it, and then uses the computer to find an optimal solution. The effect on the optimal solution of changing key elements can then be found, or alternative answers computed by systematically varying cost or requirement data. The final solution can be prepared by the computer in the form of a management report using MPS's report generator facility.

The design of MPS has been refined and improved by many years of implementation. The application of linear programming techniques accounts for a large portion of the computer time used by industrial organizations, and although this and similar languages are rather limited in the scope of the problems they can solve, their importance to the efficient operation of many large business firms cannot be overemphasized. Problems involving 100 variables and 500 inequalities (or constraints) are not uncommon. The solution of these problems "by hand" could take so long that the answer would be almost useless by the time it was obtained.

b. MATLAN

IBM's MATrix LANguage (MATLAN) is a general purpose computational system for any application that is expressible in matrix notation. It is a problem-oriented language incorporating many functional statements designed to perform such operations as matrix generation, matrix manipulation, and matrix algebra. The language can handle data in the form of scalars, matrices, or arrays, and can operate on either real or complex mathematical values.

MATLAN has been found to be a useful tool in solving mathematical problems in such areas as structural analysis, network analysis, statistics, and econometrics, and in such fields as aeronautical, civil, and electrical engineering. Specific problems to which it has been successfully addressed include systems of linear equations, ordinary and partial differential equations, integral equations, and Boolean matrix algebra.

c. PMS

IBM's Project Management System (PMS) is a set of computer program routines each of which performs a common management application function. It is a system capable of analyzing problems in the form of critical path analysis, PERT (program evaluation and review technique), PERT/

Cost, line of balance problems, life-cycle analysis, and configuration management, to mention just a few. The routines available represent some of the most advanced management techniques used by both government agencies and industry.

PMS allows for the up-to-date determination of job status and financial performance and for the investigation of probable schedule and cost implications of proposed plan changes. The techniques available with this system are useful for designating delivery dates, controlling production costs, or monitoring the use of resources such as personnel, material, money, and equipment. PMS is easy to apply and represents yet another quick and accurate use of the computer for solving everyday problems.

3. Specialized Technical Application Languages

Almost every academic discipline has been able to make use of the computer in solving its own generic problems. Toward this end, many computer languages have been developed to aid a particular functional area. Since there is such a wide variety of these languages, and since each applies to only a fairly narrow range of problems, no attempt will be made to discuss individual languages here. The reader should simply be aware that for any particular problem that requires large amounts of mathematical manipulation, there is probably a computer language or routine already designed.

In order to use these computer languages, the programmer must have a knowledge of the particular discipline for which the language was designed. Some of the areas in which languages of this type have been implemented are civil engineering, machine tool control, logical design, compiler writing, vibrational analysis, social science research, and many other minor application areas. The individual who plans to work with a specific type of problem for an extended period would find it quite worthwhile to determine whether there is an applicable specialized language and if there is, to learn that language.

D. What Language Do I Learn?

It has been stated both implicitly and explicitly throughout our discussion of computer languages that no language is best for all applications. Even though procedure-oriented languages such as PL/1 and FORTRAN offer considerably more flexibility to the user who has a deep knowledge of them, it is still difficult and time consuming to write programs in these languages to accomplish such tasks as simulation, matrix manipulation, or linear programming; the languages specifically developed for such tasks are far more easily applied. The usefulness of computer languages and their wide variety poses a problem for both professional and occasional programmers. As Naftaly [14] points out:

> Higher-level languages for business and scientific computing are firmly
> established and are gaining in popularity every day. The question is no
> longer, "Shall I use a high-level language?" but, "Which one shall I
> use?" There are indeed enough languages available [and announced]
> to make the choice a complex, if not difficult, one. Not only are there
> many languages available, but, in some cases, several dialects of one
> language exist . . . .

Allowing for all this, for what should the prospective computer user
look in selecting a language? The first consideration should be the usefulness of the language for present and future needs. What types of problem will have to be solved by computer in future work, or what current
problem could be solved on the computer presently available? This consideration of the areas of potential application is perhaps the most important selection criterion.

Another factor that should enter into the decision is the actual computing facility that is or will be available for use. If the General Electric
on-line system is to be used, then perhaps BASIC would be the logical
choice. However, if the user will have access to a large model of the IBM
System/360, then there are many procedure-oriented and application-
oriented languages from which to choose. It would be well to investigate
not only the capabilities of the computer that will be used, but also the
languages that are actually available for use on the computer. For a selected high-level language to be useful, the computer must have a compiler
to translate the coded instructions into machine language.

Closely related to the types of problem that the prospective programmer
expects to encounter is the criterion of return on investment in time spent
learning the language. The person who expects to be solving problems
of the linear or separable programming type would be wasting time by
learning a complex language such as PL/1. A few hours spent becoming
familiar with MPS would suffice for practically every problem likely to be
encountered. On the other hand, a person who expects to use the computer
in solving a variety of different scientific problems of a unique nature
would probably find that the time taken to become proficient in a language
such as FORTRAN or PL/1 would be well spent. Again, the decision depends
on the anticipated needs of the individual.

The user must also consider the future of a language: whether it will
grow to meet new computer designs and user requirements. Currently
there is about equal support for the idea of one language for all purposes
versus many languages with specific purposes. It is doubtful, however,
that any of the languages described in this chapter will become obsolete
in the near future. Of all the languages discussed, PL/1 has the best chance
of becoming a "general purpose" language (if such a language will ever

exist, and many doubt that it will). PL/1 could conceivably replace both FORTRAN and ALGOL, as well as some languages not mentioned here. But with the number of present advocates of these languages, FORTRAN in particular, the possibility of this happening soon seems unlikely. However, PL/1 is being improved and expanded periodically and might be the right choice for the user concerned with obsolescence.

Perhaps the best advice that can be given to the interested student would be to get the opinions of people who are familiar with the current state of available computer languages (computer languages are subject to great changes and new implementations are frequently announced) and then make a decision based on all of the accumulated facts. Rubey [6] sums it up as follows:

> Proper matching of a language to the job requires the separation of the facts about what is true in the real world from the often extravagant claims on the part of each language's advocates. . . . Whatever the decision, it must anticipate the future—of the chosen language itself and also of the range of uses to which it will be put. The decision thus requires a sound background of experience and a great deal of good luck.

An excellent reference for additional information concerning many of the languages presented here is that by Sammet [10]. For the purpose of actually learning a language, the reader is referred to the many texts available, as well as to the manuals distributed by such computer manufacturers as IBM and General Electric.

IV. NEW TRENDS IN COMPUTER TECHNOLOGY

It is hard to imagine a more exciting time than the present to be in the field of computer science. Only a few decades ago, the concept of the computer was little more than a curiosity for academics. Today, the study of computers is as dynamic and challenging as any field. One particularly remarkable aspect of this developing technology is that of computer hardware. The hardware market is in a great state of flux. There are computer manufacturers in direct competition with IBM for the large computer market. Hardware design is also in a period of rapid evolution. Every day, radically new components are leaving the laboratory and entering pilot production.

A. Current Evolution in Hardware Design

Hardware performance is advancing at an impressive speed. Still, most of the current changes are evolutionary, rather than revolutionary. While all of the recently introduced computers boast some combination of greater speed, smaller size, and better cost effectiveness than their predecessors, few offer any new component technology.

As we pack more and more circuitry into a silicon chip, the speed of our circuits increases for two reasons. Present-day computer circuits are so fast that the amount of time that it takes an electrical pulse to travel from one part of an integrated circuit to another is significant. Packing the circuits closer together reduces this delay and also seems to increase switching speeds. We should expect to see the trend toward higher levels of integration continue well into the 1980s.

Another important feature we can expect to see more often in the near future is the use of *pipelining,* or look-ahead processing, previously used on both the CDC Star and the Texas Instruments ASC. Pipelining is a strategy that allows execution of an instruction to begin before execution of previous instructions is completed. We can expect this concept to remain significant through 1990.

As computer processors become faster, working memories must also increase in speed in order to keep up. So far, this has been accomplished by using increasingly large-scale integrated semiconductor memory. There is reason to believe that bipolar memories will continue to evolve fast enough to keep up at least through 1990, by which time it is predicted that read/write cycle times for bipolar memories will have been reduced to 1.2 ns (1.2×10^{-9} s). Since bipolar memories require unusually large amounts of power, MOS semiconductor memories may also continue to be used and to increase in speed until they become competitive with present-day bipolar memories.

One particularly intriguing problem that will result from a growing population of increasingly quick computers is that of preparing processable data in large enough quantities to divert a data famine. The amount of data that will have to be put into these machines to utilize that increased capability far exceeds the possibility of any kind of manual keypunching. Part of the solution will be the increased development and use of direct optical character reader (OCRs) devices that can read an ordinary typed page. The U.S. Postal Service has been using such devices to read ZIP codes on envelopes for years.

B. Recent Trends

As mentioned earlier (Section I), most of the present and future advancements in computer design are and will be evolutionary. Two of the most exciting evolutions will be in the fields of mass memory design and printer design.

Two new mass memory technologies are entering the market place: magnetic bubble memories (MBMs) and charge-coupled devices (CCDs). Technically these two devices function very differently; conceptually they have much in common. Both are electronic devices that act somewhat like electromechanical rotating memory devices (such as disks and drums); that is, they can be thought of as being like carousel-type slide trays that contain binary computer data instead of slides. Both can be expected to be more reliable and easier to maintain than either disk or drums. Both will require considerably less power to operate than rotating memories and yet will be faster, although slower than semiconductor memories. Their price will be between that of rotating and semiconductor memories and will increase slowly as a function of capacity.

There are a number of important differences between MBMs and CCDs. CCDs are volatile whereas MBMs are not. This means that if the electrical power should ever be interrupted, data stored in a CCD would be destroyed, whereas data in a MBM would remain indefinitely. Thus it would not be advisable to store your system software or your highly valued data bases in a CCD because it could be destroyed, whereas on the MBM it would not. Another important difference between these two components is that MBM "rotation" can be "stopped," whereas CCDs must "rotate" continuously. This means that with a CCD the data must be read "as it goes by" and then stored in a buffer memory until the CPU is ready to use it. It is unlikely that such a buffer arrangement will be necessary for an MBM. However, CCDs may be faster by an order of magnitude than MBMs.

There is some controversy about the cost of operating a CCD memory system, with some reason to believe that a CCD mass memory could cost about one-tenth as much as a conventional MOSFET memory system. It may eventually be possible to store 400–500K on a single CCD chip. Several CCD memories are now on the market, and we can expect to see the introduction of MBM and CCD storage devices into computers within the next several years.

In the more distant future, we can expect some revolutionary changes in printer design. There is much interest in several new approaches. Perhaps the most promising ideas are ink jet, electrostatic, or laser based. Although there have been primitive demonstrations of some of these ideas, there are many unresolved problems.

C. Minicomputers, Superminicomputers, and Microcomputers

1. What Is a Minicomputer?

Defining the class of computer systems known as *minicomputers* has been a problem for years. Back in the 1960s, one attempt to delineate the minicomputer characteristics was as follows:

1. A minicomputer processor alone costs less than $25,000
2. It is employed primarily in real-time applications, such as process control

3. The vendors sell minicomputers primarily to original equipment manufacturers (OEMs) rather than to end users
4. Minicomputer architecture is characterized by a bus that interconnects the CPU, memory, and input/output channels; in mainframes, these elements are joined by point-to-point connections
5. Word sizes are 8, 12, or 16 bits
6. Address space is limited to 64K bytes (i.e., addresses are no more than 16 bits long)
7. The CPU is physically small—no bigger than a breadbox

In the 1970s, with the advent of large-scale integration (LSI) and the microprocessor, the size and price parts of these characteristics have changed dramatically. The CPU price range dropped to $2000 and $5000; the size became "smaller than a breadbox and larger than a slice of bread." The other characteristics of the mini persisted, however.

2. History of the Minicomputer

One of the first minicomputers was the PDP-5, a 12-bit computer made by the Digital Equipment Company (DEC), which hit the market in November 1963. It was replaced by the most popular family of minicomputer, the PDP-8, in April 1965. The PDP-8 with 4K words of core memory and a Teletype Model 33 ASR (when introduced) was priced at $18,000. The PDP-8 S was announced in October 1967 as a slower but cheaper version of the PDP-8, for under $10,000—the first machine at that price. Minicomputers were primarily used as OEM items in larger processing or control environments, although they had considerable appeal as stand-alone computers in a small-scale scientific environment. Typically, their limitations were as follows:

slow input and output speed of teletype-punched tape (about 5 min to load a FORTRAN compiler)

no floating-point arithmetic hardware

multiplication and division hardware available only as an option

very few available software packages

lack of a COBOL compiler

In June 1967 DEC introduced a 32K word disk for the PDP-8, which helped to overcome some of these difficulties. In November 1967 the PDP-8 I, a faster minicomputer using integrated circuits, was introduced for $12,800. This machine had a fast extended arithmetic element which could multiply in 6 μs and divide in 6.5 μs (1 μs = 10^{-6} s). DEC introduced the PDP-8/L in August 1968 for $8,500 to maintain its claim of having the cheapest minicomputer. However, in December 1968 they began to get some competition from former employees who had formed a new company, Data General, and were selling a medium-scale integrated minicomputer (16 bit), the NOVA, for $7950.

During the next few years, DEC had a great deal of competition—over 90 companies announced they were selling minicomputers—but still controlled the lion's share of the market. [DEC had never actually been alone in the minicomputer business: IBM had brought out the 1130 in February 1965 for a minimum rental of $694 per month, although they did not call it a minicomputer. The 1130 had started with the capabilities to be a business system (punched card input, line printers, and disk storage), but it was slow.] In January 1970 DEC introduced the first members of their current family of 16-bit minicomputers: the PDP-11/10 and PDP-11/20.

As of December 1977, DEC had sold over 80,000 computers and commanded 30% of the minicomputer market. Well behind DEC in revenues, but next in the ranking, come Hewlett Packard, IBM, and Data General. Data General, the company formed by former DEC employees, has sold over 40,000 minicomputers since their founding in 1969 [15].

In recent years new technology has decreased the cost of minicomputer systems while improving performance. The floppy disk or diskette introduced by IBM in 1972 has greatly decreased the cost of a disk system. Another new technology in minicomputers is the increased use of MOS (metal oxide semiconductor) memory. Bipolar-transistor-type semiconductor memories are sometimes used, but their advantage of faster speed is normally outweighed by their higher cost, larger size, and higher power demand.

3. Minicomputer Applications

Most of the currently installed minicomputers are being used in industrial control and laboratory instrumentation, the areas where it all began. The minicomputer boom started when it became apparent that the impressive recent advances in semiconductor and magnetic technologies had made it possible to construct general-purpose computers at a lower cost than the single-purpose, hard-wired controllers that were formerly used in these specialized applications. The added flexibility of stored-program computer control was a welcome bonus that helped ensure the rapid acceptance of the minicomputers.

During the past decade, the capabilities of the minicomputers have been steadily increasing while their costs have been decreasing in an equally rapid fashion. The proliferation of these small, economical, and surprisingly fast computers has led to an ever-widening range of applications for them.

Among the largest current markets for minicomputers are industrial control, research, engineering and scientific computation, business data processing, data communications, and education. Specific applications in which minicomputers are already being widely and successfully used include the following:

Process control
Numerical control of machine tools
Direct control of machines and productions lines
Automated testing and inspection
Telemetry
Data acquisition and logging
Control and analysis of laboratory experiments
Analysis and interpretation of medical tests
Traffic control
Shipboard navigation control
Message switching
Communications controllers for larger computers
Communications line concentrators
Programmable communications terminals
Peripheral controllers for larger computers
Control of multistation keys-to-tape/disk systems
Display control
Computer-aided design
Typesetting and photocomposition
Computer-assisted instruction
Engineering and scientific computations
Time-sharing computational services
Business data processing of all types

4. Superminis

Another recent arrival on the computer scene is the *supermini*—a more powerful machine that maintains the essential characteristics of the minicomputer. Some typical supermini features are as follows:

1. The CPU is still physically small but is 10 or more times as expensive as that of the ordinary mini
2. Word sizes are 24, 32, or more bits in length
3. The instruction set is upward compatible with the instructions of a family of minicomputers

4. Memory capacity is 512K bytes or more and address ranges up to 1 Gbyte (30 bits) are offered
5. Instruction execution times are faster than those for minicomputers

An example of the speed increase is given by the single-precision fixed-point add time of 500 ns for a Tandem 16 supermini compared with a conventional minicomputer execution time of 2 µs (2000 ns) or more.

More powerful instruction sets (because of a larger word size), instruction look-ahead techniques, and the use of cache memories are other features available in superminis. The *look-ahead technique* allows an instruction to be brought to the cpu from the main memory while prior instructions are being executed, so that memory access does not slow down processing. *Cache memory* is a small, very-high-speed memory used to store commonly used program memory locations to speed program execution.

Superminis were designed specifically for large scientific and business applications; in fact, some vendors now offer superminis "that replace—and the vendors claim outperform—medium scale IBM 370s" [16]. Since just about every supermini sold can be microprogrammed, very powerful instruction sets can be created inexpensively. There are even two models that emulate all the model-independent instructions of the IBM 370.

What is the difference between mainframes and superminis? Primarily, the connections among the memory, CPU, and input/output channels are made by a data bus in the supermini, whereas point-to-point connections are made in a mainframe. This means data transfers can occur simultaneously on different paths in a mainframe, but occur sequentially in a supermini.

5. *Microprocessors and Microcomputers*

In contrast to superminis, microcomputers and microprocessors (primarily eight-bit words) have been developed in the past five years. Microprocessors have all the processing logic on one LSI chip; only a few other chips, such as for clock and memory, are needed to have a computer. A microcomputer has all that is needed for a computer on one chip except a power supply. Some common microprocessor chips (2-µs cycle time) sell for under $10 in quantities of 100 or more [17]. A common microcomputer having 1 kbytes of EPROM and 64 bytes of RAM sells for under $40 in quantity [17], allowing very inexpensive computer-controlled devices to be made.

Recently, very powerful 16-bit microprocessors have been introduced. One of these is the INTEL 8086 unit which features 1-Mbyte memory address space, 8- and 16-bit operations (and allows for 8-bit 8080 software to be used with minor modifications), 24 operand addressing modes, and hardware multiplication and division.

6. The Future of Minicomputer Applications

The classical minicomputer applications, such as process control and other real-time applications, will be taken over by the dedicated microcomputer. Microcomputers will be used in many new devices, such as microwave ovens, automobiles, and in almost every computer peripheral device, as well as in very small data processing systems. Minicomputers will gradually come to be used for the same purposes as the mainframes of the early 1960s, but in much greater volume due to their lower cost.

The minicomputer growth trend is expected to continue for several years. One major reason for this is the big increase in the number of small business systems being sold. It appears that the software gap is very rapidly closing, partly due to the decrease in cost of memory, which allows inexpensive systems to use a high-level language.

REVIEW QUESTIONS

1. Discuss the major developments in the four generations of computers.

2. Explain the difference between scientific and commercial computer languages.

3. Describe the general characteristics of COBOL.

4. Summarize the advantages of PL/1.

5. What are three major reasons that PL/1 was developed?

6. What are the advantages and disadvantages of high-level languages over machine language?

7. Discuss how a professional programmer is distinguished from an occasional programmer.

8. Explain how FORTRAN was developed and how it is maintained.

9. How are programs entered and processed using APL?

10. What are the major differences between procedure-oriented and application-oriented languages?

11. Explain how computer simulation models are developed and how they differ from mathematical models.

12. List the various simulation languages and explain what type of simulation language is generally preferred.

13. What types of problem are best solved by MATLAN?

REFERENCES

1. Opler, Ascher, Effective Program Development: The Choices—Is Assembly Language Programming Passé? *Data Processing Digest* 14 (10) (October 1968).

2. McGee, William C., Effective Program Development: The Choices—"What's the Problem?" *Data Processing Digest* 14 (3) (March 1968).
3. Sammet, Jean E., *Programming Languages: History and Fundamentals*. Englewood Cliffs, New Jersey: Prentice-Hall, 1969.
4. Rosen, Saul, *Programming Systems and Languages*. New York: McGraw-Hill, 1967.
5. Rubey, Raymond J., A Comparative Evaluation of PL/1, *Datamation* 14 (12) (December 1968).
6. Rubey, Raymond J., Effective Program Development: The Choices—COBOL, PL/1, or What? *Data Processing Digest* 14 (8) (August 1968).
7. Szuprowicz, Bohdan O., The Time-Sharing Users: Who Are They? *Datamation* 15 (8) (August 1969).
8. Sharpe, William F., BASIC: *An Introduction to Computer Programming Using the* BASIC *Language*. New York: The Free Press, 1967.
9. IBM Application Program, *General Purpose Simulation System/360, Application Description*. White Plains, New York: IBM Corporation, 1967, H20-0186-2.
10. Pritsker, A. A. B., *Modeling and Analysis Using Q-GERT Networks*. New York: Halsted Press and Pritsker & Associates, Inc., 1977.
11. Pritsker, A. A. B., and Pagden, C. D., *Introduction to Simulation and* SLAM. New York: Halsted Press (Wiley), 1979.
12. Krasnow, H. S., and Merikallio, R. A., The Past, Present, and Future of General Simulation Language, *Management Science* 11 (2) (November 1964).
13. Mize, J. H., and Cox, J. G., *Essentials of Simulations*. Englewood Cliffs, New Jersey: Prentice-Hall, 1968.
14. Naftaly, Stanley M., Data Processing . . . Practically Speaking, How to Pick a Programming Language, *Data Processing Digest* 12 (11) (November 1966).
15. All About Minicomputers, *Data Processing* (December 1977).
16. Stiefel, Malcolm L., Superminis: What's in the Name? *Mini-Micro Systems* (July 1978).
17. Intel, Inc., pricelist (July 10, 1978).

RECOMMENDATIONS FOR FURTHER READING

All About Small Business Computers, *Datapro 70* (September 1978).
Applying the Minicomputer, *Data Processing Management* (1975).
Awad, Elias M., *Introduction to Computers in Business*. Englewood Cliffs, New Jersey: Prentice-Hall, 1977.
Bairstow, Jeffrey N., Mr. Iverson's Language and How It Grew, *Computer Decisions* 1 (1) (September 1969).
BASIC *Language Reference Manual*. General Electric Company, Information Service Department, May 1967.
Branscomb, Lewis M., Promising Areas of Research in the Computer Industry, *Physics Today*, pp. 55–61 (January 1976). This excellent article seems to be intended as inspiration for the young physicist looking for career direction. It does a commendable job of presenting some of the highlights of what is going on in the field and should be fairly readable for the nonphysicist as well.

Buxton, J. N., *Simulation Programming Languages*. Amsterdam: North-Holland, 1967.

Crouch, Harry R., Cornett, John B., Jr., and Eward, Ronald S., CCDs in Memory Systems Move into Sight, *Computer Design*, pp. 75–80 (September 1976). This should be fairly useful to the nonengineer.

Emshoff, J. R., and Sisson, R. L., *Design and Use of Computer Simulation Models*. New York: Macmillan, 1970.

Fenves, Steven J., Problem-Oriented Languages for Man–Machine Communication in Engineering, *Proceedings of the IBM Scientific Computing Symposium on Man–Machine Communication*. White Plains, New York: IBM Corporation, Data Processing Division, 1966.

Gordon, Geoffrey, Simulation Languages for Discrete Systems, *Proceedings of the IBM Scientific Computing Symposium on Simulation Models and Gaming*. White Plains, New York: IBM Corporation, Data Processing Division, 1966.

Gordon, Geoffrey, *System Simulation*. Englewood Cliffs, New Jersey: Prentice-Hall, 1969.

Gould, R. L., GPSS/360—An Improved General Purpose Simulator, *IBM Systems Journal* 8 (1) (1969).

Herzog, B., Mechanical Vibration Simulation: Past and Future, *Proceedings of the IBM Scientific Computing Symposium on Digital Simulation of Continuous Systems*. White Plains, New York: IBM Corporation, Data Processing Division, 1967.

Hill, Fredrick J., and Peterson, Gerald R., *Digital Systems: Hardware Organization and Design*. New York: Wiley, 1973. A very well-written text book on hardware design.

Hoff, George, System-Level Integration Shrinks Size and Cost of Medium-Scale Computer, *Computer Design*, pp. 81–88 (April 1976). This very good article not only discusses circuit integration, but also general design trends, particularly in regard to communications and channel design. The author is the Engineering Manager for Systems Development in DEC's Large Computer Group. By coincidence, most of his examples deal with the then new Dec-system-20. The article is very readable but should be read with caution—it might convince you to buy a DEC computer.

IBM Application Program, *Mathematical Programming System/360, Application Description* (H20-0136-3). White Plains, New York: IBM Corporation, 1968.

IBM Application Program, *Project Management System/360, (360A-CP-04X) Version 2, Application Description Manual* (H20-0210-1). White Plains, New York: IBM Corporation, 1968.

IBM Application Program, *System/360 Continuous System Modeling Program (360A-CX;16X), Application Description* (H20-0240-2). White Plains, New York: IBM Corporation, 1968.

IBM Application Program, *Project Management System/360, (360A-CP-04X) Version 2, Program Description and Operations Manual* (H20-0344-2). White Plains, New York: IBM Corporation, 1968.

IBM Application Program, *System/360 Continuous System Modeling Program (360A-CX-16X), User's Manual* (H20-0367-2). White Plains, New York: IBM Corporation, 1968.

IBM Application Program, *Mathematical Programming System/360 (360A-CO-14X), Version 2, Linear and Separable Programming—User's Manual* (H20-0476-1). White Plains, New York: IBM Corporation, 1968.

IBM Application Program, *System/360 Matrix Language* (MATLAN), *Application Description* (H20-0479-1). White Plains, New York. IBM Corporation, 1968.

IBM Application Program, *MARVEL/360 (360A-CO-15X) Primer* (H20-0496-0). White Plains, New York: IBM Corporation, 1968.

IBM Application Program *MARVEL/360 (360A-CO-15X) Program Description Manual* (H20-0505-0). White Plains, New York: IBM Corporation, 1968.

IBM Application Program, *System/360 Matrix Language* (MATLAN) *Program Description Manual* (H20-0564-0). White Plains, New Yprk: IBM Corporation, 1968.

IBM Application Program, *Project Management System/360 (360A-CP-04X) Report Processor System Manual* (Y20-0085-1). White Plains, New York: IBM Corporation, 1967.

Introduction to Programming in BASIC. General Electric Company, Information Systems Division, January, 1967.

Itel to Offer Nat'l cPUs; Eye 148/158, *Electronic News,* pp. 1–50 (11 October 1976). A very good description of the Itel announcement with a reasonable level of detail. While *Electronic News* is one of the more useful trade periodicals, it may be difficult to find. The only library in the state of Oklahoma that carries it is the Tulsa City County Library, 200 Civic Center, 4th and Denver, Tulsa. Unfortunately, they only keep the most recent six months. There are two libraries in Texas with certain back issues. They are located at North Texas State and Southern Methodist University.

Itel's Powerful New Computer, *Business Week,* pp. 74–76 (25 October 1976). A good analysis of Itel's merchandise, strategy, and prospective buyers.

Johnson, Bruce B., Requirements for Real-Time Digital Simulation Systems, *Proceedings of the IBM Scientific Computing Symposium on Digital Simulation of Continuous Systems.* White Plains, New York: IBM Corporation, Data Processing Division, 1967.

Kenney, Donald P., *Minicomputers.* American Management Associations, 1973.

Klein, Stanly, The Minicomputer Boom in 1978, *Mini–Micro Systems* (July 1978).

Kolsky, H. G., Problem Formulation Using APL, *IBM Systems Journal* 8 (3) (1969).

LSI Circuit Technology Cuts Computer Size/Cost, Ups Speed 100%, *Computer Design,* pp. 27–28 (January 1976). A fair description of the Amdahl 470V/6, its highlights, and their significance.

McLean, Joe, DEC Unveils System in the IBM 370 Range, *Electronic News,* p. 1 (19 January 1976). A fairly good introduction and description of the then new Decsystem-20.

Medium Scale Systems, *Datamation,* p. 134 (August 1976). Announces and describes the replacement of IBM's models 370/145 and 370/135 by the 370/148 and 370/138.

Naylor, T. H., and Balintfy, J. L., Burdick, D. S., and Chu, K., *Computer Simulation Techniques.* New York: Wiley, 1966.

Operating Systems Extended, *IBM Computing Report,* p, 7 (Summer 1972). This issue contains product announcements for much of IBM's current line. The

OSU library files it (incorrectly) with their current periodicals on the first floor.

An Overview of Minicomputers, *Data Processing Management*. Pennsauken, New Jersey: Auerbach, 1975. A reference service notebook, continuously updated.

PL/1 at the Crossroads, *Datamation* 14 (12) (December 1968).

Rideout, V. C., Continuous Systems Simulation—Present Programs and Future Possibilities, *Proceedings of the IBM Scientific Computing Symposium on Digital Simulation of Continuous Systems*. White Plains, New York: IBM Corporation, Data Processing Division, 1967.

Sammet, J. E., Problem In, and a Pragmatic Approach To, Programming Language Measurement. Fall Joint Computer Conference, 1971.

Sammet, J. E., An Overview of Programming Languages for Special Application Areas. Spring Joint Computer Conference, 1972.

Schlesinger, Stewart, On-Line Engineering Analysis—Present and Future, *Proceedings of the IBM Scientific Computing Symposium on Digital Simulation of Continuous Systems*. White Plains, New York: IBM Corporation, Data Processing Division, 1967.

Second Conference on Applications of Simulation, *Datamation* 15 (2) (February 1969).

Stiefel, Malcolm L., Superminis: What's in the Name? *Mini-Micro Systems* (July 1978).

Surden, Esther, Itel Brings Out Two CPU's with IBM Compatibility, *Computerworld*, pp. 1–7 (18 October 1976). Good coverage of the announcement of Itels AS/4 and AS/5.

Taub, A. H., On Time-Sharing Systems, Design and Use, *Proceedings of the IBM Scientific Computing Symposium on Man-Machine Communications*. White Plains, New York: IBM Corporation, Data Processing Division, 1966.

Teichroew, D., and Lubin, J., Computer Simulation—Discussion of Technique and Comparison of Language, *Communication of the Association for Computing Machinery* 9 (10), 723–741 (October 1966).

Thompson, F. B., The Future of Specialized Language. Spring Joint Computer Conference, 1972.

Turn, Rein, *Computers in the 1980s*. New York: Columbia University Press, 1974. The author of this intriguing document is a senior systems analyst for the Rand Corporation, which specializes in military research contracts. The author spends much time on his approach to forecasting, on general purpose civilian computers, on military command control computers, and on the ability of computer components to survive the effects of nuclear radiation. There apparently is a good deal of research going on in this area. The development of a nuclear-effects-resistant computer could do much to improve the credibility of America's nuclear arsenal. Currently, we have the capability to destroy this planet about 17 times. The development of an appropriate computer could solve the problem of who pushes the button the 2nd–17th times.

User Ratings of Minicomputers and Small Business Computers, *Datapro 70* (November 1977).

Chapter 6
The Management Science Process:
A Systematic Approach

This chapter is directed toward the development of a comprehensive and practical procedure for the determination of the appropriate management science technique for the solution of models developed to solve various management science problems. It presents the structural elements of such a procedure, contains a brief description of each solution method, and describes the various parametric considerations. These elements are of necessity interrelated by virtue of the model parameters. The chapter begins with the general problem type (by content and by form) and then develops the general model. The first major categorization deals with situations of certainty versus those of uncertainty. Problems of uncertainty are further categorized into detailed discussions of the elements of deterministic and stochastic situations. Problem symptomology is related in brief discussions of the many solution methods that exist for both deterministic and stochastic problem situations. For more details about specific techniques mentioned in this chapter, the reader is referred to the books available in the operations research libraries. The list of books and articles recommended for further reading at the end of this chapter includes some of the most comprehensive publications to accommodate the reader's needs.

I. APPROACHING SOLUTION METHOD SELECTION

A managerial problem does not arrive with a tag that notes its catalog number, handbook table, or the name of the systems analysis algorithm qualified to accomplish its resolution. In fact, problems do not generally arise in such manner that clearly indicates their contents, scope, and type. They must therefore be classified, both by content and by form. As mentioned earlier (Chapter 2, Section I.A), the problem content classifications may well parallel the major elements of the organization, i.e., research and development, production, marketing, finance, accounting, industrial relations, and logistic applications. The problem whose effect and structure overlaps one or more of these areas must be analyzed with respect to the strategic elements thereof. Such an analysis allows the problem to be broken into manageable segments.

The conversion to classification by form generally casts problems as one or more of the following types: inventory, allocation, forecasting, queuing, routing, sequencing, replacement, competition, and search (see Chapter 2, Section I.B).

Following classification by form, the analyst would define the variables of the problem (Chapter 2, Section I.C) and develop a preliminary model. The models encountered in this type of analysis are usually symbolic or mathematical (Chapter 2, Section II). All models must periodically be checked throughout the solution method selection (and even application) process to provide the best possible "fit." This is accomplished through *analysis of model parameters* and subsequent *model refinement* (see Figure 1), both of which will be discussed in more detail with reference to particular problem classifications (see Sections II.A and III.A). Such model revision prevents errors in the choice of solution method, thereby reducing solution time and effort.

At this point many problems can be resolved or the method of their solution can be decided with certainty (though possibly incorrectly). The acceptance or rejection of contract bids is an example of a *definite solution method* (see Figure 2). If we assume the low bidder to have submitted a proposal which meets carefully established specifications within a pre-

FIGURE 1. Model revision.

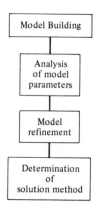

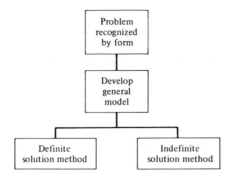

FIGURE 2. Definite vs indefinite solution methods.

determined and acceptable cost and service range, the decision (problem resolution) is essentially obvious.

On the other hand, there are problems, such as those dealing with an old but carefully designed set of inventory policies, one or more of which have slowly gone sour, where the solution is not obvious although numerous symptoms exist and probably have existed for some time. These symptoms are much like those experienced by an individual who has gradually become overweight or a heavy smoker. They are bearable because the situation has deteriorated slowly and the system has attempted to compensate. Response to perturbations occurs, but not properly or in a timely fashion. The precise solution to such a problem is not obvious as is the acceptability of contractual bids noted earlier. This uncertainty as to how to resolve the situation requires the use of an *indefinite solution method*.

At this point, problems can be classified as deterministic or stochastic (probabilistic) (see Figure 3). While many situations exhibit characteristics of both types, an intertwining of the two approaches usually occurs at only the most detailed levels of the solution.

FIGURE 3. Deterministic vs stochastic problems.

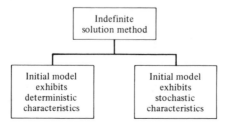

II. DETERMINISTIC PROBLEMS

As mentioned in Part I, a solution can be said to be *deterministic* if the parameters (constants or elements whose values characterize one or more of the variables entering into the situation) of the problem take on unique values. If, for example, the problem concerns determination of the number of tires used to support a standard American passenger car (before it has been modified for use as a dune buggy or other nonoriginal purpose), the answer, four, is deterministic.

While early problem models may be felt to contain characteristics that indicate the strong possibility of reaching a unique solution, deterministic situations split into two complementary subsets: determinate (unique solution) problems and optimization problems (see Figure 4).

A. Determinate Problems: Model Revision

Determinate problems are those with models wherein the number of unknown variables exactly equals the number of equations; this yields a unique solution to the problem. Before choosing a solution method for a determinate problem, however, the model may require some revision:

Analysis of model parameters: Detailed analysis is needed of the parameters contained in the initial model. Are these the correct ones? Are they the only ones? Are all of them necessary? Other considerations include the interrelationships among the elements of the model, the availability of data, and the correlation among the problem, the model, and the environment of the problem. Perhaps some other approach may be better suited to the situation or other pertinent elements are missing. Care must be exercised to ensure that the model fits the situation.

Model refinement: As more thought and effort is given to the problem and the experience of additional personnel is obtained, the model will most likely begin to change. Some parameters will be found to be inappropriate. Holes that require filling will be noted. This cyclic analysis of the situation (see Figure 5) will yield a much firmer model.

FIGURE 4. Determinate vs optimization problems.

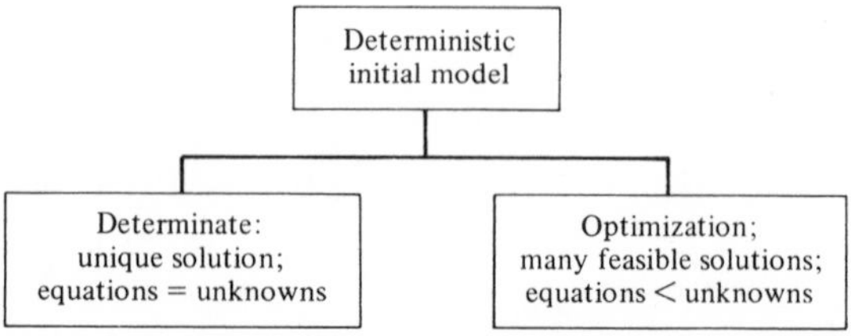

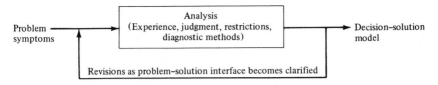

FIGURE 5. Firming the decision model.

During this period of analysis, as the relationships between the problem symptoms and possible methods of solution become fixed, it may even become apparent that the problem type does not actually yield a unique solution but rather requires optimization. Assuming, however, that the refined model is still determinate in nature, an appropriate analytical solution method should be chosen.

B. Analytical Solution Methods

Although definitely relied upon during the phases of data collection and data conversion for solving stochastic and optimalistic problems, analytical methods by themselves can only be used for the solution of determinate problems. These generally fall into the classifications of economic studies, deterministic networks, and certain production planning and line balancing problems (see Figure 6).

1. Economic Studies

The solution of problems and the implementation of resulting decisions require the utilization of capital. The efficiency of the utilization of that capital is dependent upon a number of factors—some technical, some

FIGURE 6. Applications for analytical solution methods.

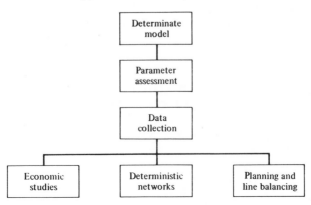

monetary, and some nonmonetary. Economic studies are for the purpose of determining whether, or in what manner, a problem solution should be undertaken or modified to obtain the best use of available capital and to take into account all pertinent factors. The general objective is thus to permit the *maximization of financial efficiency*.

Economic studies include the final stages of analysis for the economic comparison among alternatives (as noted previously), the calculation of facility and equipment depreciation rates, lot sizes for manufacture or purchase, and break-even analysis.

a. Compound Interest Problems

Practically all problems can be solved in more than one way. Both costs and results must be evaluated when choosing between the alternatives. Capital not used for a proposed venture may be utilized in some other manner, even if only left in the bank. Thus money also has a time value; a sum of money is worth more next year than it is today by an amount equal to the interest it could earn if invested. The free enterprise system is one in which capital or material worth is put to work to earn more capital. If capital is allowed to remain dormant, no risk is taken and no money is earned, except for inflation and devaluation effects.

Because of the time value of money and the fact that implemented problem solutions themselves have lifespans, certain monetary analysis is necessary when evaluating alternative solutions. Several widely accepted methods may be used to make this evaluation. The most common methods are those of determining the *equivalent annual cost* (see Figure 7) or the *present worth* of the alternatives. Either method results in the removal of the time as a factor in the evaluation of alternatives. In addition to the appropriate interest model (possibly modified to fit a given situation), values for certain parameters are required:

1. The interest rate
2. The expected project life
3. The initial expenditure
4. Expected costs during the project life and when they will occur
5. A salvage value, if any

b. Depreciation Calculations

The calculation of equipment and facility depreciation is necessary when determining the total worth of a given enterprise or the amount of taxes that either are due and payable or may occur during the life of a project. Several methods are accepted and available for this task (see Figure 8). Among these, the following are most popular:

Straight line: An equivalent amount each year for the life of the item,

$$(\text{Initial item value—Expected salvage})/n$$

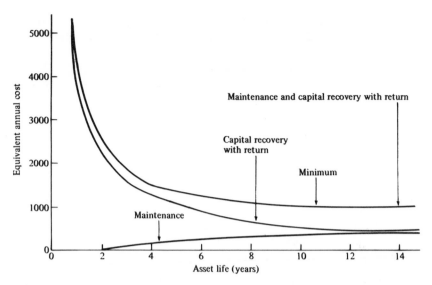

FIGURE 7. Graphical presentation of the expected history of an asset. *From Thuesen [1], p. 178.*

FIGURE 8. Comparison of values obtained by various depreciation formulas. *From DeGarmo [2], p. 90. (Copyright© 1967 by Macmillan Publishing Co., Inc.)*

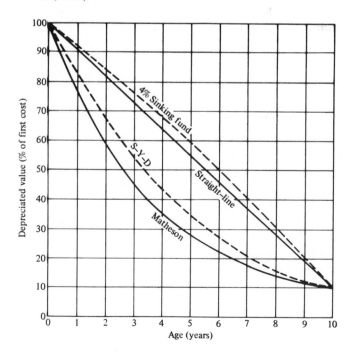

Sum of the digits: An accelerated (or reduced) rate wherein a percentage is calculated annually based on the number of years of life remaining (or past) divided by a number that represents the sum of the years of item life. For example, the accelerated depreciation for the second year of life of an item having a 5-year life would be

$$\frac{4}{1 + 2 + 3 + 4 + 5} = \frac{4}{15} \cong 27\%$$

whereas the reduced rate would be

$$\frac{2}{15} \cong 13\%$$

Either rate would permit the entire amount to be taken into account during the item's life, but the former permits a faster depreciation during the early period, whereas the latter permits a slower depreciation during the same period.

Declining balance: A fixed percentage of the remaining book value of the item for the preceding year is used for the annual depreciation charge. A variation of this, called the *double-declining balance*, permits a much faster depreciation during the early portion of the life of the item.

In addition to the applicable depreciation model, values for several parameters are required:

1. The expected item life
2. The expected salvage value
3. The initial investment in the item
4. An appropriate interest model (only if an alternative method, the *sinking fund*, is chosen)

c. Deterministic Inventory Models

Concern for economic lot sizes exists when attention is directed toward inventories of raw materials and finished goods. Inventory is taken to compensate for the lack of synchronization between the supplier and the production (or selling) process, among the various stages within the production process, and between the production process and the demand for the product. The maintenance of inventory also provides for safety stock to compensate for uncertain demand and the economic considerations of quantity discounts, economic order (or manufacturing lot size) quantities, and production smoothing. Inventory is expensive. There are costs associated with ordering material (or for tooling up for production), its value, transport, and storage [taxes, obsolescence, spoilage, utilities, and protection (both from the elements and from people)], and loss of interest that could have been obtained had the capital been invested elsewhere.

Sound management requires the lowest possible total inventory expense. The establishment of sound inventory policies requires the knowledge of how much to make (or buy) and when to make (or buy) it. A partial solution to these questions is obtained by means of one or more of the stochastic forecasting techniques, as material use is almost always at a varying rate. When reliable knowledge about quantity is available, analytical techniques can be used to determine the lowest-cost method for providing the necessary inventory. This is accomplished by finding that quantity level where the costs of the ordering process (or the task of tooling up), the material cost (including any possible discount order quantities), and the holding cost yield the lowest possible total cost (see Figure 9):

$$\text{Total cost} = \text{Order (or production) cost}$$
$$+ \text{ Material cost} + \text{ Holding cost}$$

Using the following parameters this general model is solved analytically to find the minimum:

1. The various elements of the cost of holding stock
2. Any discount order quantity provisions
3. The demand pattern
4. The maximum production rate

FIGURE 9. Deterministic inventory model: economic order quantity determination wherein annual demand is 1500 units, ordering cost is $8.33 per order, and storage cost is $0.10 per unit per year.

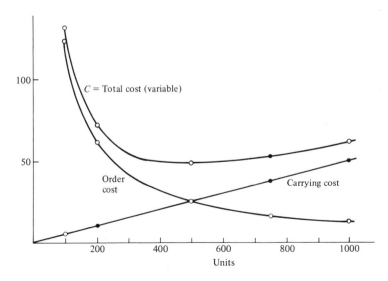

5. The cost of the ordering process
6. The cost of tooling up
7. The desired economic quantity

d. Break-Even Analysis

It is a common experience that the costs of obtaining or operating a facility or item of equipment are affected by one or more variables which may be defined and analyzed. This leads to a type of economy study that is of considerable importance in the selection of alternatives or the operation of equipment or facilities in such a manner as to obtain maximum economy. When a common variable affects the costs of two methods, there may be a certain value of the variable for which the costs for both methods will be equal. This value of the variable is called the *break-even point*. Many cases involving break-even points can be expressed by mathematical equations and the break-even point found by their solution. In other situations, conditions may be such that the functional relationships are not continuous and, therefore, not directly solvable. A graphical approach called a *break-even chart* is then useful.

When comparing alternative solutions to a problem, break-even analysis may yield an unsuspected item of information: At different levels, first one and then the other solution may be the more economical. Such analysis may illustrate the sensitivity of the estimated level of operation and indicate the need for additional data collection.

One of the most satisfactory uses of break-even charts is to show the relationship between income and costs of a business. The balance sheet and income statement give information about a business for one particular time or period but do not show what would happen to the profits if the rate of production changed. From the standpoint of analysis and control, it is of considerable advantage to be able to depict the effect of a change in the production rate (assuming sufficient demand to absorb a higher production). Such break-even analysis is therefore highly useful when attempting to determine the feasibility of a new product.

2. Deterministic Networks

Many work projects comprise a vast number of interrelated steps. This type of work is usually characterized by a strong interdependency among parts of the total job; that is, some parts cannot be started until other parts are finished. When there are thousands of such parts making up the total job, it becomes very difficult for management to perform its normal functions, such as assigning work at specific times, assigning personnel to specific jobs, and so on. The job frequently becomes even more complicated when tight schedules must be met. Fitting together many thousands of pieces of a complicated puzzle and making everything flow smoothly is a formidable task even for the most competent manager.

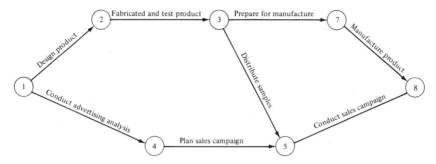

FIGURE 10. Arrow diagram for sample deterministic network problem (see Table 1).

Work projects having a known series of steps, a known relationship among the various steps, and a similarity to earlier projects (so that resource estimates can be made with a degree of certainty) can be depicted in the form of a *deterministic network*. The existence of such a network and attendant resource estimates permits the use of an analytical technique known as the *critical path method* (CPM) (see Section III.B.4). Use of CPM facilitates the determination of the expected project duration time, permits the scheduling of personnel, materials, and equipment, and indicates bottlenecks. Should unforeseen problems arise, reanalysis can be performed to reschedule in an optimal manner, possibly making up for lost time. The work schedule resulting from use of the method provides an excellent basis for noting work progress. Parametric values needed for CPM are as follows:

1. A knowledge of the predecessor–successor relationships among all project steps (see Table 1, Figure 10)
2. The minimum direct cost (resource) estimate to complete each step
3. A time estimate to complete each step at minimum cost
4. An estimate of the minimum time to complete each step

TABLE 1. Estimated Values for Sample Deterministic Network Problem

Predecessor	Successor	Activity description	Normal Time	Normal Cost	Crash Time	Crash Cost
1	2	Design product	6	2000	2	4000
1	4	Conduct advertising analysis	4	1000	1	5000
2	3	Fabricate and test product	12	2000	3	4000
3	5	Distribute samples	9	1500	2	3000
3	7	Prepare for manufacture	6	1000	4	4000
4	5	Plan sales campaign	10	1250	2	4500
5	8	Conduct sales campaign	15	6000	7	10000
7	8	Manufacture product	6	2500	2	5125

5. An estimate of the direct cost to complete each step in the minimum feasible time (see Figure 11)
6. The per-time-period indirect cost to complete the project

A CPM total cost solution curve is shown in Figure 12.

3. Planning and Line Balancing

One of several heuristic methods such as Gantt charts, the indicator method, or the ranked positional weight technique is applicable to certain production planning and production line balancing problems.

a. Gantt Charts

Gantt charts are relatively simple horizontal bar-graph-type charts which can be used to display a planned work schedule, depict the jobs competing for given equipment items during a specific period of time, or store/display tentative decisions while a set of assignments is manually determined. A Gantt chart can also be utilized to depict accomplishments versus plans. Such charts (and accompanying analysis) are quite useful in job shop situations and wherever extensive project networks are not involved.

FIGURE 11. Direct cost–time relationship, the basis of the CPM solution method.

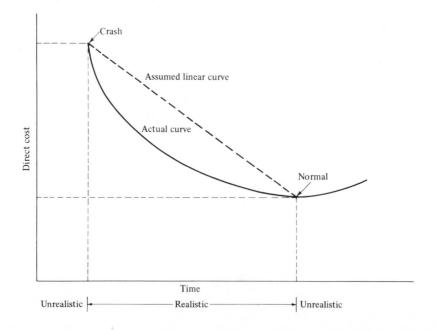

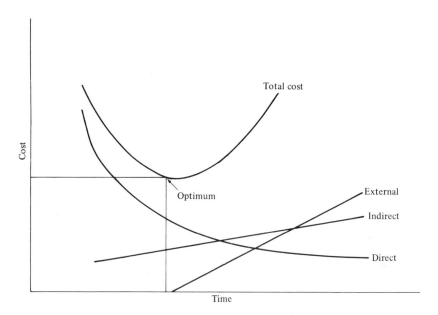

FIGURE 12. CPM total cost solution curve.

Parametric considerations include:

1. Resources of concern (personnel, equipment, and/or facilities)
2. Time requirements
3. Minor sequencing concerns

When planning job assignments with the aid of a Gantt chart, several different schemes are possible for resolving conflicting needs for resources. Among these are selection on the basis of first-come–first-served, shortest operation first, or random selection decision rules. None of these methods is of an optimizing nature, although any will produce a feasible plan. The "goodness" of the resulting plan is highly dependent upon the skill of the planner.

b. Indicator Method

A heuristic work assignment scheme known as the *indicator method* is sometimes useful for scheduling job shop operations. This method consists of identifying all equipment items (or crews, facilities) that can accomplish the necessary operations. The effectiveness of the assignment(s) of the task to each of the units of equipment is determined by noting the amount of machine time that would be required for each assignment. The indicator is found by computing the ratio of job time on each equipment item to the least amount of time required on any piece of equipment.

Assignments are then made to that available piece of equipment having the lowest indicator value. Such a scheme is useful when new jobs must be added to an existing work plan without major revisions. The parametric concerns include:

1. Time requirements for each machine type
2. Existing work assignments

c. Line Balancing

Production line balancing, the task of attempting to assign equivalent work loads to each station along a production line, is generally a sticky problem. A conglomeration of individual tasks that must be accomplished in a somewhat fixed sequence (e.g., painting cannot be done before welding), does not usually permit the design of work stations each having an equal standard or average time requirement (see Figure 13). However, it is preferable that the best balance possible be established. One heuristic method useful for this task is the *ranked positional weight technique*. This method functions by using a set of decision rules to rank and combine job elements into reasonably balanced operations for assignment to the various work stations. Ranking is based on the sum of sequence times for all elements that must be subsequent to the element under consideration; the greater the sum, the higher the rank (see Table 2). Grouping of elements into operations is based on the cycle time for the line, the subsequent sequence time (which takes care of most of the technological ordering requirements), and any remaining spurious order conflicts (Table 3). Parametric considerations are as follows:

1. The individual element times
2. The cycle time
3. The technological order constraints (depicted in a network)

Note that although the resultant groupings are feasible, they are *not necessarily optimal*.

TABLE 2. Ranking of Elements Shown in Figure 13

Element	Subsequent sequence time	Immediately preceding elements	Element	Subsequent sequence time	Immediately preceding elements
01	235		06	128	05
03	214		08	128	07
04	214		09	106	
02	192	01	12	106	06, 08
10	192	03, 04	13	84	09, 12
11	170	02, 10	14	84	11
07	165		15	63	13, 14
05	161		16	0	15

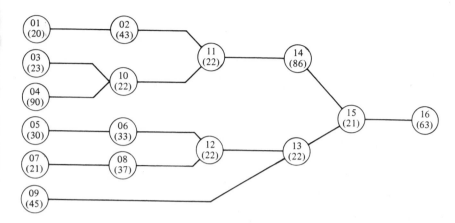

FIGURE 13. Sequence of elements to be ranked and grouped by the ranked positional weight method. Element durations (say, in minutes) are
given in parentheses. Total element time: 600.

C. Optimization Problems: Model Revision

Optimization problems are significantly different from determinate ones
in that there are a number of feasible (workable) solutions and more unknowns in the situation than there are equations or other means for evaluation. Optimization solution techniques permit the best (optimal) feasible
solution to be selected. As in the determinate case, however, some revision of the initial problem model may be necessary (refer to Figures 1
and 5).

Analysis of model parameters: The first step is an analysis of the significance of each parameter with regard to the total model environment.
Each parameter should be both necessary and correct with respect to its
application within the model. At this point it should also be determined
which are the *strategic* parameters—those on which the successful so-

TABLE 3. Resultant Grouping of Elements Shown in Figure 13

Operation	Elements assigned	Time assigned
1	01, 03, 02	20 + 23 + 43 = 86
2	04	90
3	10, 11, 07	22 + 22 + 21 = 65
4	05, 06	30 + 33 = 63
5	08, 09	37 + 45 = 82
6	12, 13	22 + 22 = 44
7	14	86
8	15, 16	84

lution of the model will depend. The form that these strategic parameters will assume will, in large measure, determine the management science technique to be used in the solution of the model. Due to the interactive nature of parameters in the optimization problem situation, this analysis must remain as only a preliminary one.

Model Refinement: The refinement phase of the model revision procedure centers around the firm establishment of those elements that will lead to a solution fulfilling the initial intent of the model. Cyclical analysis often exposes gaps which must be filled by the addition of new parameters or by the expansion of those already in the model. It is during this period that the true nature of the model under analysis emerges. It may be found that the model is not of the optimization class but rather has a unique solution, or perhaps calls for a stochastic solution method. Should such a situation occur, a new analysis with respect to the relationship between the model and the new problem class must be undertaken. If the results of this analysis indicate that the model should indeed be solved by optimization methods, the next step is to determine the specific method to use.

D. Optimization Solution Methods

Optimization methods cover a wide range of specific solution techniques which may be applied independently or in conjunction with other techniques of either the same or a different nature. Optimization methods are extremely useful in the solution of alternative strategy or decision problems such as transportation, allocation, and other associated problems. Most of the optimization techniques fall into the category of mathematical programming, although auxiliary techniques such as exhaustive enumeration, network analysis, and iterative techniques may also be applied in optimization (see Figure 14).

The determining factor in selecting a solution technique is the nature of the parameters involved in the model. These parameters must reflect both the intent of the model and also the solution adaptability of the chosen technique. The following sections consider some of the available optimization techniques and their application requirements.

1. Mathematical Programming

The application of *mathematical programming techniques* yields valid representations of real-world situations only to the extent that the initial models represent such validity. Optimization problems can be and often are classified by their attributes: linear or nonlinear, continuous or discrete, static or dynamic, deterministic or probabilistic (the last of which,

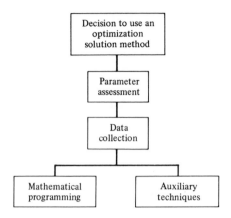

FIGURE 14. Use of optimization solution methods.

of course, requires stochastic techniques, which are considered in Section III). Mathematical programming techniques are built upon three main ingredients:

1. *Choice variables:* Those variables whose values are manipulated in the search for a best solution. These variables reflect the decision choices available to the manager.
2. *Constraints:* Relationships among variables that restrict the values assignable to the choice variables.
3. *Objective function:* A mathematical expression that defines the optimization process (in terms of maximization or minimization) and whose value may be computed when the values of all variables are specified.

Selection of an appropriate optimization solution technique is dependent upon the nature of these ingredients (see Figure 15).

a. Linear Programming

Linear programming is the process of optimizing a linear objective function subject to a linear set of constraints, where *linearity* involves the absence of exponents (no power series). A model, developed from a real-world situation, must be fitted to one of the standard linear programming forms.

Linear-based situations are extremely common in the field of management and linear programming has therefore often been utilized. Problems of optimal mix among production processes and optimal loading of those production processes represent two of the types of models that may be optimized through the use of linear programming. Due to the single-level nature of the variables involved in the linear program, solutions can usu-

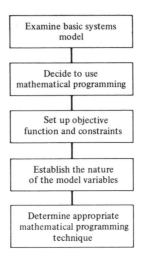

FIGURE 15. Model evaluation for optimization.

ally be readily obtained. Problems involving a large number of variables, however, may require computer solutions for feasibility (see Figure 16).

Simplex and Special Models

The most common linear programming solution technique is the *simplex algorithm*. This is a computational algorithm that derives a solution by moving from one feasible solution to another along the lines representing the constraint functions until the optimum is found. Three special cases of linear programming are transportation, transshipment, and job assignment.

Transportation: In its obvious sense, the *transportation model* seeks the minimization of the cost of transporting a certain commodity from a number of sources to a number of destinations. Although transportation problems can be solved using the regular simplex method, the specialized model offers a more convenient procedure through the use of some shortcuts, yielding a less cumbersome computational scheme.

Transshipment: Situations arise in which, because it may not be economical to ship directly from sources to destinations, a commodity may pass through one or more of the other sources and destinations before eventually reaching its ultimate destination. This case is referred to as *transshipment*. Although the transportation model cannot be used directly to handle this problem, a slight modification could allow its use.

Job assignment: The assignment of different jobs on different machines comprises a linear optimization problem when the objective is to minimize total cost (given that different assignments incur different costs). The formulation of such a problem may be regarded as a special case of the transportation model, where the jobs represent ''sources'' and the machines represent ''destinations.''

Parametric Programming

Often problems arise where it is desired to optimize with respect to a number of different objectives having relative priorities. *Parametric programming* provides the analyst with a technique to handle such problems. This method is a direct outgrowth of linear programming and allows the analyst to examine the extent to which the optimal solution may be modified without affecting the validity of the objective function (i.e., without adversely affecting the optimal solution).

Integer Linear Programming

When dealing with a linear programming problem in which all variables must be integers, the intersections of the constraints will not specify feasible solutions as they do in the standard linear programming problem. The linear programming problem is thus concerned with *continuous* func-

FIGURE 16. Flow diagram of computerized linear programming system. *Courtesy of International Business Machines Corporation.*

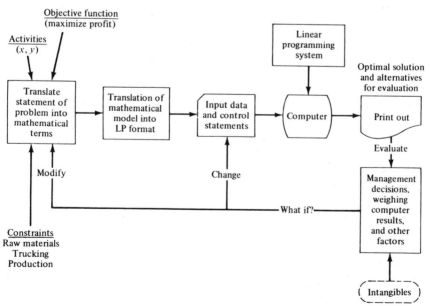

tions, whereas integer linear programming is concerned with *discrete* functions. When all variables in an otherwise linear programming problem are constrained to integer values, an *integer linear programming* problem is involved. When some, but not all, variables are so constrained, a *mixed integer problem* is involved. When all variables are constrained to take on values of either zero or one, we have a *zero–one* programming problem (see also Binary Programming). Problems of these types are commonly encountered in decision analysis.

The general solution procedure is to start from a basic solution that does not satisfy the integral constraints of the problem, then add an equation chosen such that all integral programs necessarily satisfy it. The introduction of such an equation results in the appearance of a new variable having a negative value and, consequently, permits the application of the dual simplex algorithm. If an integral solution to the original problem is still not found, a new equation is added and the process is iterated.

b. Nonlinear Programming

Nonlinear programming is a general form of solution technique that may be used with models whose objective functions and constraints assume any level of nonlinear form (i.e., includes exponential mathematics). The problem faced by the analyst in formulating the nonlinear problem is similar to that for the linear problem, but finding a solution is considerably more complicated.

The formulation for the nonlinear programming inputs involves a determination of both the objective function and the environmental constraints on the problem. The nonlinear programming problem allows for the establishment of a higher level of abstraction with respect to the elements upon which the model is formulated, and special techniques are required to solve it since the feasible solution regions are not as well behaved as in the linear case. The simplex procedure in a maximization problem would move from one extreme point to another until movement in any direction decreased the value of the objective function. Although this would guarantee an optimal solution in the linear case, in the nonlinear case it may lead to a *local* rather than a *global* or absolute maximum. For this reason, the following special programming techniques have been tailored to the specific types of nonlinear problem.

Convex Programming

The optimum of a nonlinear programming problem will in general not be at an extreme point of the constraint set and may not even be on its boundary. Also, the problem may have local optima, as distinct from a global optimum. These properties are direct consequences of the problem's nonlinearity. However, it is possible to define a class of nonlinear problems, called *convex programming* problems, that are guaranteed to be free of distinct local optima. A convex programming problem involves the minimization of a convex (upward-curved) objective function [or the maximization of a *concave* (downward-curved) objective function] over a convex constraint set.

Geometric Programming

This mathematical programming technique is based on the geometric inequality of positive numbers or variables; that is, the geometric mean of two or more positive numbers is always less than or equal to their arithmetic mean. This technique deals with both constrained and unconstrained minimization problems; however, it can be applied to maximizing functions given the appropriate changes in its inequalities.

Separable Programming

This technique deals with nonlinear problems in which the objective function and all constraints are expressable as the sum of some single-variable functions. The approximate solution can be obtained for any separable problem using piecewise linear approximation and the simplex method of linear programming. A mixed integer procedure can then be applied to ensure that the given approximation is valid.

Quadratic Programming

This is another outgrowth of the basic nonlinear programming technique; its application is primarily limited to those cases where the objective function is primarily of quadratic form but the constraints are linear. Several solution procedures have been developed for this particular case because of its frequent application in management and scientific areas.

Discrete Programming

When all variables of a nonlinear programming problem are constrained to have, say, integral values, the problem is one of *discrete programming*. The discrete programming problem allows the decision variables to assume any of a finite set of values specified by the problem constraint equations. Because of the complexity involved in the solution of the discrete programming problem in its initial form, this type of problem is most often converted to a binary type for solution. The conversion involves a restructuring of the objective function constraints in such a manner that the variables are restricted to zero and one in the finite solution set.

Binary Programming

When all variables in a problem are constrained to have values of zero or one only, the problem is said to be of a *zero–one* or *binary* type (see Integer Linear Programming). Binary programming is valuable for two reasons. First, many management problems can be easily portrayed as zero–one problems (when "either/or" constraints and/or variables are involved). Second, nonlinear integer and discrete programming problems may be restructured such that they become binary in nature. This restructured problem provides the analyst with a tool which can be more readily applied to complex problems encountered in industrial applications. The simplification to binary form expands the size of the problem that may be undertaken.

2. Auxiliary Techniques

a. Exhaustive Enumeration

Given any problem having a finite number of alternative solutions, one method useful in locating the optimal solution is *exhaustive enumeration*. This is simply the listing of all solutions prior to the choice of the optimum. This is the oldest form of optimization aid and may still be applied in many situations where the number of alternatives is manageable. High-speed digital computers have greatly increased the size of the problems that may be handled by this method.

b. Network Analysis

Branch and Bound

The most widely adopted approach for solving general integer programming problems, the *branch and bound* technique, uses a method of tree search, sometimes referred to as a *backtrack algorithm*. The consistency of this technique with computer logics provides the possibility of computer application over a wide range of management problems. Several versions of the approach are developed for large-scale computers and have been applied successfully on decision models.

The strength of the branch and bound technique lies in its ability to reduce the number of alternatives that must be evaluated. Although originally applied to the transportation problem (see p. 142), this technique has also been applied to sequencing, allocation, traveling-salesman, and quadratic-level control problems, as well as others.

Dynamic Programming

Dynamic programming is an approach to dealing with multistage decision processes, that is, problems that involve a sequence of decisions. The objective function and constraint equations considered under this technique may take on almost any form: linear, nonlinear, tabular, and so on. Their form will affect the complexity of the solution process but should not affect the theoretical relationships involved in the dynamic programming problem.

The relationships required for the application of dynamic programming center on its interstage nature. The inputs to a dynamic programming problem include the input to each stage, the stage decision, the associated stage return, and the stage output. The constraint relationship that interconnects the various stages of the problem are known as the *problem states*. The state relationships provide the transition from decisions made at each stage in the chain. The most common aspect of the dynamic programming problem is the nature of the solution technique involved—generally, a recursive optimization approach. However, in recent years several alternatives have been applied to the basic approach, and final-value and decision-based solutions are now available to the analyst.

c. Iterative Techniques

Heuristic Programming

A heuristic decision process suggests one in which an attempt is made to determine not the optimal decision per se, but rather the best decision for the available time. A *heuristic program* is a computational procedure that, when appropriately applied, will yield a good solution in a finite number of steps. The basic heuristic program consists of a set of procedures or rules to be applied to various problem situations in order to obtain an acceptable solution. For example, for project scheduling the procedure might be that the project having the least total slack would be processed first, with the other projects then being scheduled in ascending order of available total slack. Note that such a procedure would lead to a good schedule but *not an optimal one*.

Search Techniques

These techniques are widely applied on unconstrained problems (or constrained problems converted to unconstrained form). There might be many variables involved, but the technique, generally, searches for an improvement in objective function by changing the value of a few variables at a time along an appropriate direction and by an appropriate increment or decrement. The example of a blind person searching for the top of a hill by taking small steps uphill or downhill can best represent the concept of these techniques. Many sophisticated procedures have been developed for various sorts of function, and digital computers have facilitated the cumbersome computations involved in these techniques.

Mathematical programming and associated techniques are summarized in Table 4 and Figure 17.

III. STOCHASTIC PROBLEMS

The word *stochastic* comes from the Greek στοχαα̧εθαι (*stochazethai*) meaning to aim or guess. A stochastic process is one wherein the parameters can take varying values. The solution of a stochastic model refers to probable outcomes: "on the average" or with a certain probability a certain thing or value will occur. A common example is a weather forecast, which is now usually given with an occurrence probability. (A radio announcer might well read a forecast predicting a "40% chance of rain" even though a heavy downpour is in progress.)

A. Model Revision

As with models for deterministic problems, a certain amount of revision is necessary to ensure significant solutions.

Analysis of model parameters: The analysis of parameter values associated with stochastic problems assumes the added dimension of the probabilistic influence. The analysis must center around two main

TABLE 4. Mathematical Programming Parameter Summary

Technique	Parameter relationship
Linear	Linear objective function
	Linear constraints
Parametric	Linear program (plus sensitivity analysis)
Integer	Variables must assume integer values
	Linear functions
Nonlinear	Nonlinear objective function and/or
	Nonlinear constraints
Convex	Convex objective function (min.) and convex solution space
	or
	Concave objective function (max.)
Geometric	Nonlinear functions of special form
Separable	Functions expressable as single-variable functions
Quadratic	Quadratic objective function
	Linear constraints
Discrete	Integer nonlinear
Binary	Variables must assume values of zero or one
Auxiliary techniques	
Exhaustive enumeration	Any kind of function
	Manageable number of variables
Branch and bound	Computational algorithm
Dynamic	Staged decision process
	Nonlinear or linear functions
Heuristic	Yields nonoptimal but good solution
	Limited resources

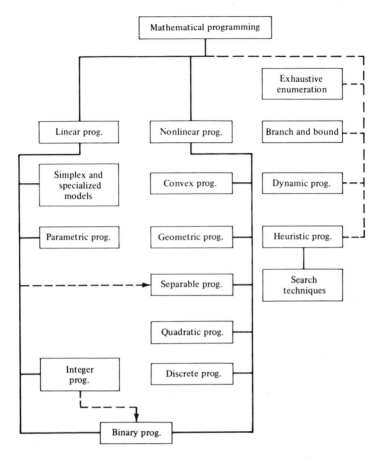

FIGURE 17. Mathematical programming and associated techniques.

factors: the objective of the preliminary model and the nature of the variables involved in that model. The objective of the model will lead the analyst to the general area of solution techniques. However, the specific solution technique chosen will depend in large part on the variables associated with the individual problem. If the variables can readily be related to the situation expressed in a given solution technique, then that technique may be applied to the model under consideration. However, if the variables do not fit any standard solution technique, then the analyst is faced with a model redefinition problem (see Figures 1 and 5).

Model refinement: The model refinement process may assume two basic forms under the stochastic problem situation. First, if the model variables do not fit any known solution technique, then the model must either be expanded or contracted to bring the variables into line. Second, if the variables do fit a standard solution technique, then redefinition will center around the accomplishment of the objectives of the model. If the solution results fail to achieve this stated purpose, then the model must be altered in such a manner that the desired results may be obtained.

B. Stochastic Solution Methods

The techniques described here for solving stochastic problems do not represent an exhaustive listing but rather illustrate the potential areas available to the management analyst (see Figure 18).

1. Forecasting and Prediction

The determination of a future outcome on the basis of past data is the fundamental purpose of many management models. Forecasting and prediction are two types of technique that may be directly applied to the

FIGURE 18. Stochastic problem solution methods (partial list).

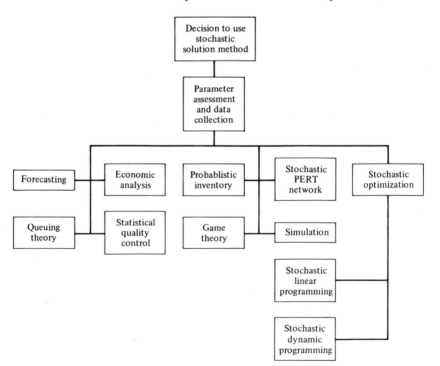

problem of decision on future action. The difference between the two techniques lies in the nature of the data that forms the basis for the analysis. Under *forecasting* the estimation of the occurrence of a future event is based on previous data generated by the event in question. Under *prediction*, the estimation is based either on data generated by similar elements or similar environmental conditions or on subjective estimates.

Whether forecasting or prediction is chosen will depend on the individual problem situation. The amount of data available will in turn restrict the type of subtechnique used to produce the estimation. These subtechniques are many and are applicable to both forecasting and prediction.

Regression analysis, one of the most widely known techniques, is concerned with the fitting of a line or curve to a given set of points. Linear regression, the most common form of application, establishes a relationship in the form of the equation of a straight line with a defined slope. This equation can then be utilized to determine further points in relation to the desired future event.

Another common method of predicting future occurrences is the *moving average*. This is calculated by taking averages of groups of data points in each of *n* time periods, thus yielding a series of time interval plots. The moving average produces a curve that reflects the underlying progress of the system while minimizing its random fluctuations. The longer the period of consideration for a moving average, the more the fluctuation will be reduced and the smoother will be the curve formed by the moving average. Since for too long a period there is a tendency for the moving average to iron out the underlying variations, particularly where it is markedly nonlinear, it is best to use as short a period as possible consistent with a reasonably smooth moving average [3, pp. 6–7].

A modification of the moving average technique is the *weighted moving average*, which differs in that the most recent values are adjusted so as to carry more weight in the predicted value. Both techniques are to be plotted at the midpoint of each period; thus it is convenient to base the average on an odd number of time increments so that the midpoint of the total period coincides with an actual increment.

Another technique that is often applied in the area of forecasting and prediction is *exponential smoothing*, a method of removing random fluctuations in the data from which the prediction is drawn. A smoothing coefficient is applied to the error between the actual and predicted values of the data set in order to obtain new predicted values.

2. Stochastic Economic Analysis

Economic studies were discussed in Section II.B.1 from a deterministic standpoint; such words as "expected," "distribution," and "average" did enter the discussion, however. From a general viewpoint, many reasonably good solutions can be obtained by ignoring the stochastic aspects of the real world. This is especially true as concerns depreciation calculations (since rate of item use rarely enters the calculation) and compound interest calculations (since identical stochastic processes would apply to all alternatives being compared).

In most real-life cases, an economic study is undertaken due to several probabilistic factors. For example, probabilistic cash flows are to be established in most economic studies, such as calculation of the minimum annual revenue requirement. Probabilistic capital budgeting is another pertinent area.

One important application of probability theory in engineering economy is in estimating the *expected value* of certain kinds of extreme event. Expected values are frequently used in the comparison of alternative investments.

In studying the entire economy of sociopolitical systems where the stochastic variables of the macrodynamic system (e.g., national income, national product, total employment) are involved, deterministic methods are ineffective; therefore stochastic analysis is largely applied.

3. Probabilistic Inventory Models

The variability found in real-world situations should be considered when performing break-even analysis. Situations dealing with economic (manufacture or purchase) quantities (i.e., inventory) form a subset of break-even analysis and therefore may undergo stochastic analysis. Parametric concerns for general stochastic break-even analysis depend upon the given situation; the remainder of this section deals with stochastic economic quantity analysis in particular.

Deterministic economic quantity analysis (Section II.B.1.c) assumes perfectly linear rates of material use and either the instantaneous receipt of ordered material or an absolute unvarying period of time between order and receipt (called *lead time*). *Stochastic economic quantity analysis* considers real-world variability and "plans" for it by including what is known as a *safety stock* when determining an inventory policy. This safety stock is determined to allow for differences between actual and planned use rates, production rates, and lead times. Rather involved statistical computation is necessary to provide both safety from outages caused by the worst combination of all factors and relief from the expense caused by attempting to insure against ever having an outage. Policies are then formed by combining the results of the safety stock calculations with the minimum total cost calculations discussed in Section II.B.1.c. Parametric considerations include:

1. The various elements of the cost of holding stock
2. Any discount order quantity provisions

3. The demand pattern
4. The maximum production rate
5. The cost of the ordering process
6. The cost of tooling up
7. The desired economic quantity

4. Program Evaluation and Review Technique (PERT)

As noted earlier in the discussion of deterministic networks (critical path method—Section II.B.2), many work projects comprise a vast number of interrelated steps. The CPM discussion was directed at situations where estimates of resource requirements for each project step could be reliably obtained from experience with similar or related work projects. Not all projects, however, are similar to earlier work. Many organized activities occur at or near the frontiers of existing knowledge and technology and as such cannot readily be likened to other tasks for the purpose of generating reasonably precise estimates of resource needs (especially calendar time) for the completion of the project. On the other hand, initial planning for such projects does include a delineation of major project steps and their interrelationships. The *program evaluation and review technique (PERT)* was designed to aid in the management of such projects. With CPM, resource needs and time requirements are treated as deterministic. PERT methods treat these parameters as random variables from probability distributions unique to each major step (usually referred to as activities). The initial version of PERT was structured so as to treat the time requirement as stochastic for a given set of values for the personnel and funding parameters. Subsequent versions of PERT have been devised to permit resources other than time to be treated as stochastic.

Typical PERT approaches require three estimates for the amount of time required for each activity:

1. *Optimistic time:* The time duration under the most favorable conditions
2. *Most likely time:* The time duration under normal conditions
3. *Pessimistic time:* The time duration under the least favorable conditions

$$\text{Expected} = \frac{\text{Optimistic} + 4(\text{most likely}) + \text{Pessimistic}}{6}$$

These time estimates for each activity are used as parameters of a β-distribution for that activity and are used to determine both the expected time and the variance for the project step. From this point, routine statistical analysis can be used to find the *critical path*—that series of activities which establishes the expected time of completion of the entire project. Management can then concentrate its efforts on the small portion of the entire project that most affects the project completion. In addition, many "what if" questions can be answered, such as, "What if more talent is assigned to work on a particular task?" The overall effect of such an assignment can be readily determined; if beneficial, it can be implemented.

As in the case of CPM, rescheduling of work segments and reassignment of resources can be accomplished when unforeseen events cause disruptions in the initial planning or generate an urgency for completion. Probable effects due to changes in the level or direction of effort in any segment of the project can readily be noted. Parametric values needed to use this method are as follows:

1. A knowledge of the predecessor–successor relationships among all project steps
2. An estimate of the optimistic duration of each project step
3. An estimate of the most likely duration of each project step
4. An estimate of the pessimistic duration of each project step

5. Queuing Theory

Queuing theory deals with the analysis of waiting lines, such as job shop work-piece flow, trucks or ships waiting to be unloaded, and aircraft waiting to land. It is amazing how many situations can be formulated as waiting-line problems. Even so, it is useful to examine the aspects of many such problems so as to construct a simulation model in order to obtain very close approximations in specific cases. At this point, a number of queuing terms need definition.

Unit: An object that arrives at the queue for service. It may or may not have to wait for service.

Service: That which is performed on the units.

Service facility: The place at which the units arrive and at which the service is rendered. There can be one or more service facilities. Facilities are sometimes called *channels* or *stations*.

Service policy: The specifications as to exactly what the service includes.

Queue: The waiting line; the units waiting to be served. There can be one or more queues in the situation being analyzed.

Input: The units arriving for service.

Output: The units leaving the facility after servicing has been completed.

Queue discipline: The order in which the units in the queue are serviced. This is usually on a first-come–first-served basis, but there may be priority interrupts.

System: The combination of facility(s) and queue(s).

The interval of time between successive arrivals is rarely constant; rather, it is usually a random variable. Similarly, the amount of time required to perform the service on a unit is also usually a random variable. The distributions of these random variables are determined from empirical observations.

There is no limit to the number of different queuing situations that might arise. For example, some units will leave the line due to impatience, and there could be switching from one queue to another. A unique queuing model must be constructed for each different situation.

The purpose of studying waiting-line situations is to find ways of improving existing systems (e.g., by eliminating bottlenecks and/or reducing costs) or to design proposed new systems optimally. For example, with a given arrival distribution and service-time distribution, it is possible to determine the appropriate number of dock units to put at a truck terminal. In the study of a time-sharing computer system, it is possible to determine at what point (in terms of received calls) another telephone line should be added to handle terminal users. Waiting-line analysis typically involves the determination of the distributions and expected values of the following random values:

1. Queue length
2. Length of nonempty queues
3. Number of units in the system
4. Waiting time of an arrival
5. Time an arrival spends in the system

In addition to the system configuration, parametric concerns for making a queuing study include:

1. The arrival time distribution
2. The queue discipline (e.g., first-come–first-served, priority, random selection)
3. The service time distribution for each facility

6. Statistical Quality Control

Statistical quality control is useful not only at various points in the production process, but also as a means for assuring the suitability of raw materials and purchased parts acquired from vendors. Statistical quality control was given a firm start during the early part of World War II. Prior to that time, the control of product quality was maintained by partial or total inspection of finished units and by inspection of items initially produced by a new setup of a machine or a line.

Comparative studies made by firms that adopted the methods of statistical quality control showed the new method to produce significant savings in scrap, rework, and even salary (due to reductions in inspection personnel time). Such savings are possible because periodic samplings of units at various points in the production process enable management to detect the point where suitable relations to tolerance begin to become marginal and drift into an unsuitable relationship. Corrective action can then be taken before other than random occurrences cause units to become little more than scrap or to require expensive rework. Likewise, the complete inspection of all finished units is not necessary. The maintenance of acceptable tolerances at all the stages of manufacture ensures the suitability of the finished unit without the need and expense of checking prior to shipment.

Along a similar line, statistical sampling of incoming raw materials and purchased parts ensures that only acceptable items enter the production process. Similar assurance could be obtained through total inspection of incoming material, but such a procedure would be expensive. Furthermore, in many cases it is not possible; for example, chemical content cannot be determined without destroying the usability of the material.

Statistical quality control is based on the laws of probability. Such laws hold that the characteristics of a suitable sample of a population reliably match the characteristics of that population. The trick is the acquisition of a suitable sample. A series of proven procedures permit this acquisition with little difficulty.

Parametric concerns are dependent upon the process being monitored. For example, somewhat divergent methods of sampling exist for processes where a number of defects can occur per unit (e.g., defects per yard of cloth) without causing the product to become useless and processes where the occurrence of a single defect renders the unit useless (e.g., a dry cell with a botched electrode).

One of the tools important to statistical quality control is the *control chart*. A control chart is a statistical device principally used for the study and control of repetitive processes. It may be used first to define the goal or standard for a process that the management might strive to attain, then to help attain that goal, and finally to judge whether the goal has been reached. It is thus an instrument to be used in specification, production, and inspection, uniting these three phases of industry into an interdependent whole.

7. Game Theory

Competitive situations exist whenever two or more persons have objectives and courses of action such that an increase in the chances of obtaining one person's objectives reduces the chances that others will attain theirs. There are numerous examples of such adversarial relationships: parlor games, military battles, political campaigns, and the advertising and marketing campaigns of competing business firms. A basic feature in most of these situations is that the final outcome depends primarily on the combination of strategies that are selected by the adversaries. Game theory (Chapter 3, Section V) is a mathematical theory that has been developed to describe how competitors should behave in such circumstances. It places particular emphasis on the decision-making process of the competitors in that each attempts to rationalize the behavior of the others in an attempt to act optimally.

The bulk of the research on game theory has been on two-person zero-sum games—those that involve only two competing units and whose sum of the net winnings is zero. In analyzing such a game, it is assumed that each player wishes to maximize the minimum amount of guaranteed personal gain, no matter what the other player does (see the discussion of the maximin principle in Chapter 3, Section IV.B). When each player is faced with a finite number of courses of action, the problem can always be solved by allowing mixed strategies in which each course of action is chosen by a chance device with predetermined probabilities. Such stochastic decision rules ensure that the players are unable to take advantage of any prediction that they may make concerning the behavior of the opponent.

The basic limitation of game theory is the inability of the players to ascertain accurate values for the win–loss result for any combination of strategies played. It is not difficult to establish that one outcome is preferable to another, but it is quite another thing to state the precise degree of preferability. The primary usefulness of game theory is that it provides a conceptual framework within which to consider a competitive problem. Actions and the opponent's probable counteraction can be considered. Parametric needs are as follows:

1. A listing of all possible actions by the player
2. A listing of all possible actions by the opponent
3. The consequence of each combination of actions

In today's management applications game theory has been incorporated into multiobjective programming.

8. Simulation

Simulation is the general-purpose tool of management science and has no doubt been used for more than any other tool or technique when no analytical technique can be developed for solving the given problem. Much of the usefulness of simulation is due to the recent advances in the field of high-speed digital computers, allowing the analyst to perform thousands of simulation runs in a relatively short period of time. (See Chapter 2, Section IV.D and Chapter 5, Section III.C.1.)

Modern management simulation is based on a mathematical model of the system under analysis. Rather than directly describing the overall behavior of the system, the simulation model describes the operation of the system in terms of single events of the individual components of the system. The system is divided into elements whose behavior can be assumed, in terms of probability distributions, for each of the possible states of the system and its inputs. The interrelationships between the elements are also built into the model. Thus, simulation provides a means of dividing the model-building job into smaller component parts and then combining these parts in their natural order, allowing the computer to present the effect of the interaction among them.

After constructing the model, it is activated in order to simulate the actual operation of the system and record its aggregate behavior. By repeating this for the various alternatives and comparing their performance, the analyst can identify the best alternative. Because of statistical error, it is impossible to guarantee that the alternative yielding the best simulated performance is indeed the optimal one; however, it should be at least near optimal under the given assumptions of the simulation model.

The most common method of simulation is known as the *Monte Carlo method*. This technique involves the setting up of a statistical distribution based on past data. A random number generator is then used to indicate the occurrence of the simulated event, with the random numbers weighted by the statistical distribution. The simulation of a number of random number draws is a predictor of future performance.

Although Monte Carlo simulation is by far the most well known, in the past few years considerable effort has been expended in the construction of special, direct simulation techniques. These direct simulations usually involve preconstructed computer programs. One of the better known of these is GPSS (Chapter 5, Section III.C.1.a). The direct simulation programs generate an artificial sample of arrivals or service times. More recent computer simulation languages are explained in Chapter 5.

The basic purpose of most simulation studies is to compare alternative courses of action. Thus, the simulation program must be flexible enough to accommodate readily the alternatives that will be considered. Therefore, it is most important that the strategy of the simulation study be planned carefully before finishing the simulation program.

9. Stochastic Optimization

Deterministic optimization techniques (Section II.D) are inapplicable to stochastic problems, but there are several methods of stochastic optimization, two of which are introduced here.

a. Stochastic Linear Programming

This technique deals with situations where some or all the parameters of the problem are described by random variables rather than by deterministic quantities. Such cases seem typical of real-life problems where it is difficult to determine the values of the required parameters exactly. In the case of linear programming, sensitivity analysis can be used to study the effect of changes in the problem's parameters on the optimal solution. This, however, represents a partial answer to the problem especially when the parameters are acutely random variables. The objective of stochastic programming is to consider these random effects explicitly in the solution of the model.

The basic idea of all stochastic programming models is to cast the probabilistic nature of the problem in the form of an equivalent deterministic model. One of the techniques used most often is *chance-constrained programming*.

b. Stochastic Dynamic Programming

This technique is concerned with problems that involve a sequence of stochastic decision processes. Proper structuring of the model containing random elements must take into account the intermingled sequence of decisions and emerging information about the exact values of random elements.

Most stochastic dynamic programming problems are not much more difficult to solve than their corresponding forms. Once familiar with deterministic dynamic programming, one can handle probabilistic versions of the problem by considering the following questions:

1. What is an optimal policy for a deterministic version of the model?
2. How much information about the probability distribution does an optimal solution require?
3. How does the dynamic programming algorithm for the stochastic model differ from that for the deterministic version?

REVIEW QUESTIONS

1. Contrast the differences between definite and indefinite solution methods.

2. What are the differences between deterministic and stochastic models?

3. Discuss how a Gantt chart can be used in planning, scheduling, and control.

4. Explain how the indicator method is used in scheduling job shop operations.

5. State some problem areas in management where you think integer programming can be useful.

6. Distinguish between transportation and transshipment models.

7. Describe the concept of convex programming.

8. Under what circumstances would you prefer to use forecasting techniques rather than prediction?

9. Explain the key features of forecasting models.

10. What are the parametric considerations of probabilistic inventory models?

11. Discuss different time estimations required in PERT.

12. What is a critical path?

13. Outline the model-building process that is used to develop a dynamic programming model.

14. Give four specific examples of cases where statistical quality control might be applicable.

REFERENCES

1. Thuesen, H., *Engineering Economy*, 4th ed. Englewood Cliffs, New Jersey: Prentice-Hall, 1975.
2. DeGarmo, E. Paul, *Engineering Economy*, 4th ed. New York: Macmillan, 1972.
3. Gregg, J., *Mathematical Trend Curves: An Aid to Forecasting*. London: Oliver and Boyd, 1964.

RECOMMENDATIONS FOR FURTHER READING

Text Books

Avriel, Mordecai, *Nonlinear Programming: Analysis and Methods*. Englewood Cliffs, New Jersey: Prentice-Hall, 1976.
Bradley, Stephen P., Hax, Arnoldo C., and Magnanti, Thomas L., *Applied Mathematical Programming*. Reading, Massachusetts: Addison-Welsey, 1977.
Cooper, Robert B., *Introduction to Queueing Theory*, 2nd ed. New York: North Holland, 1981.
Garfinkel, Robert S., and Nemhauser, George L., *Integer Programming*. New York: Wiley, 1972.
Johnson, L., and Montgomery, Douglas C., *Operations Research in Production Planning, Scheduling, and Inventory Control*. New York: Wiley, 1974.
Kleinrock, Leonard, *Queueing Systems*. New York: Wiley, 1975.
Lasdon, L., *Optimization Theory for Large Systems*. New York: McMillan, 1970.
Orlicky, Joseph, *Material Requirements Planning*. New York: McGraw-Hill, 1975.
Polak, E., *Computational Methods in Optimization*. New York: Academic Press, 1971.
Simmons, Donald M., *Linear Programming for Operations Research*. San Francisco: Holden-Day, 1972.
Taha, H. A., *Operations Research: An Introduction*. New York: McMillan, 1971.
R. Thierauf and R. Grosse, *Decision Making Through Operations Research*. New York: Wiley, 1970.

Wagner, Harvey M., *Principles of Operations Research*. Englewood Cliffs, New Jersey: Prentice-Hall, 1969.

Zionts, Stanley, *Linear and Integer Programming*. Englewood Cliffs, New Jersey: Prentice-Hall, 1974.

Articles

Bartholdi, John J., III and Ratliff, H. Donald, Unnetworks, with Applications to Idle Time Scheduling, *Management Science* 24 (8), 850–858 (April 1978).

Chelst, Kenneth, An Algorithm for Deploying a Crime Directed (Tactical) Patrol Force, *Management Science* 24 (12), 1314–1327 (August 1978).

Christensen, John, and Obel, Borge, Simulation of Decentralized Planning in Two Danish Organizations Using Linear Programming Decomposition, *Management Science* 24 (15), 1658–1667 (November 1978).

Collins, M., Cooper, L., Helgason, R., et al., Solving the Pipe Network Analysis Problem Using Optimization Techniques, *Management Science* 24 (7), 747–760 (March 1978).

Darden, William R., and Perreault, William D., Jr., Classification for Market Segmentation: An Improved Linear Model for Solving Problems of Arbitrary Origin, *Management Science* 24 (3), 259–271 (November 1977).

Davignon, Gilles R., and Disney, Ralph L., Queues with Instantaneous Feedback, *Management Science* 24 (2), 168–180 (October 1977).

Drezner, Z., and Wesolowsky, G. O., A Trajectory Method for the Optimization of the Multi-Facility Location Problem with LP Distances, *Management Science* 24 (14), 1507–1514 (October 1978).

Fischer, M. J., An Approximation to Queueing Systems with Interruptions, *Management Science* 24 (3), 338–344 (November 1977).

Fishman, George S., Grouping Observations in Digital Simulation, *Management Science* 24 (5), 510–521 (January 1978).

Gorry, G. Anthony, et al., Relaxation Methods for Pure and Mixed Integer Programming Problems, *Management Science* 18 (5), 229 (January 1972, part I).

Jagannathan, R., A Minimax Ordering Policy for the Infinite Stage Dynamic Inventory Problem, *Management Science* 24 (11), 1138–1149 (July 1978).

Klein, Dieter, and Holm, Soren, Integer Programming Post-Optimal Analysis with Cutting Planes, *Management Science* 25 (1), 64–72 (January 1979).

Kleindorfer, Paul, and Kunreuther, Howard, Stochastic Horizons for the Aggregate Planning Problem, *Management Science* 24 (5), 485–497 (January 1978).

Kurgackner, M. G., Production Smoothing Under Piecewise Concave Costs, Capacity Constraints, and Nondecreasing Requirements, *Management Science* 24 (3), 302–311 (November 1977).

Liittschwager, J. M., and Wang, O., Integer Programming Solution of a Classification Problem, *Management Science* 24 (14), 1515–1525 (October 1978).

Lucas, William, An Overview of the Mathematical Theory of Games, *Management Science* 18 (5), 3 (January 1972, part II).

McDaniel, Dale, and Devine, Mike, A Modified Benders' Partitioning Algorithm for Mixed Integer Programming, *Management Science* 24 (3), 312–319 (November 1977).

Morton, Thomas E., An Improved Algorithm for the Stationary Cost Dynamic Lot Size Model with Backlogging, *Management Science* 24 (8), 869–873 (April 1978).

Smith, Donald R., Optimal Repairman Allocation—Asymptotic Results, *Management Science* 24 (6), 665–674 (February 1978).

Talbot, F. Brian, and Patterson, James H., An Efficient Integer Programming Algorithm with Network Cuts for Solving Resource-Constrained Scheduling Problems, *Management Science* 24 (11), 1163–1174 (July 1978).

Winston, Wayne, Optimal Assignment of Customers in a Two Server Congestion System with No Waiting Room, *Management Science* 24 (6), 702–705 (February 1978).

Wood, Steven D., and Steece, Bert M., Forecasting the Product of Two Time Series with a Linear Asymmetric Error Cost Function, *Management Science* 24 (6), 690–701 (February 1978).

Zuller, Klaus, Deterministic Multi-Term Inventory Systems with Limited Capacity, *Management Science* 24 (4), 451–455 (December 1977).

PART II
MANAGEMENT SCIENCE APPLICATIONS

This part of the book discusses how the management science tools and techniques described in Part I are actually applied in different areas. Rather than attempting to present a complete picture of the entire spectrum, it covers only the most familiar and better known tools and techniques in any detail.

Table 1 is presented to provide both the novice and the experienced management scientist with a handy reference source of the current tools and techniques that may be applied to certain functional areas. This table is by no means a total view of the capabilities of these tools and techniques in aiding decision making. The problem areas listed are by necessity very general. Specific adaptations to other areas are certainly possible and, in fact, desirable, for that is the singular beauty of all the techniques—they breed new and broader applications and advanced, more efficient methods. The user-selected methods and their applications are included in the body of the book. Readers desiring an in-depth study should refer to the References and Recommendations for Further Reading in Chapters 7 and 8.

Table 2 is provided as a reminder of the various tools needed for successful practice of management science techniques.

Dearden [1] classified managerial information systems into major systems and minor systems. A *major system* is one that affects the entire structure of an organization, that is, financial, personnel, and logistics systems. A minor system is one that is confined to a single functional part of an organization, such as research and strategic planning.

Chapter 7 is devoted to the major management science applications in the areas of finance and accounting, personnel, and logistics. Logistics,

TABLE 1. Application of Management Science Techniques to MIS Subsystems

Management science techniques	Marketing and planning	R&D and engineering	Manufacturing and quality control	Purchasing inventory and distribution	Labor and industrial relations	Finance and accounting	Total firm
Linear programming	X	X	X	X	X	X	
Economic order quantity				X			
Economic lot quantity			X				
Break-even analysis	X					X	
Improvement curve		X	X			X	
CPM network analysis	X	X	X	X	X	X	X
Time-series analysis	X				X	X	
Dynamic programming	X		X			X	
Dimensional analysis	X	X					
Symbolic logic		X			X		

Functional areas

D
E
T
E
R
M
I
N
I
S
T
I
C

165

STOCHASTIC						
Heuristic modeling	X					
Sensitivity analysis	X	X				
Decision theory	X					X
Competitive modeling	X					X
Queuing theory	X	X	X	X		
Statistical quality control		X	X			
PERT network analysis	X	X	X	X	X	X
Monte Carlo analysis	X	X	X			
Behavioral modeling			X	X		
Markov process	X				X	
Simulation	X	X	X	X	X	X

TABLE 2. Summary of Application of Mathematical, Probabilistic, and Statistical Tools to Management Science Techniques

| | | Techniques | | | | | | | | | | | | | | | |
| | | Deterministic | | | | | | Stochastic | | | | | | | | | |
Tools		Linear programming	Dynamic programming	Economic order quantity	Economic lot quantity	CPM	Decision theory	Competitive modeling	Game theory	Simulation	Forecasting	Sensitivity analysis	Queuing	PERT	Economic analysis	Statistical quality control	Heurestic modeling
Mathematical	Calculus	×	×	×	×		×	×	×	×		×	×		×	×	×
	Matrix analysis	×	×			×	×	×	×								
	Weighted average							×		×	×					×	
	Moving average										×					×	
	Smoothing										×						
	Theory of inequality	×	×									×					

Statistical and probabilistic	Bayes' theorem						X	X					X	X
	Conditional probability						X	X				X	X	X
	Probability distribution				X				X	X	X	X	X	X
	Monte Carlo theory				X		X		X	X	X	X		X
	Random number generation				X		X		X	X	X	X	X	X
	Expected values			X	X		X	X	X	X	X	X	X	
	Utility theory				X		X	X	X		X			
	Sampling							X	X			X		
	Regression analysis								X	X	X	X		X
	Correlation analysis					X			X	X		X	X	X
	Statistical test								X	X	X		X	
	Time series						X					X		
	Markov chain						X							
Others	Parameter analysis	X												
	Break-even analysis						X				X			
	Network analysis		X							X				
	Graphic analysis	X	X							X	X			
	Reliability	X	X				X			X				X

in this context, include systems related to material flow within an organization, such as production, inventory, and transportation.

Chapter 8 covers the minor management science applications with special emphasis on the area of research and development.

REFERENCE

1. Dearden, J., *Managing Computer-Based Information Systems*. Homewood, Illinois: Irwin, 1971.

Chapter 7
Major Management Science Applications

The business environment of today is subject to more changes than ever before. Management therefore is faced with the responsibility of making decisions of major importance that will to some extent affect every functional unit of the organization. The old management tools of intuition, personal experience, and understanding can no longer be trusted as the sole basis for the best possible decision. However, as we have seen, scientific processes are increasingly being used to expand the manager's decision-making capacity with additional information and alternative courses of action.

In this chapter we introduce the application of management science tools and techniques to the fields of accounting and finance, personnel, production, inventory, purchasing, and marketing. Our purpose is only to provide some background; the discussion is not intended to be exhaustive.

I. ACCOUNTING AND FINANCE

Accounting and finance are two of the main areas of concern within most organizational efforts. Each is dependent on the other for valuable information for the decision process. The main objective of *finance* is to set up fiscal policies or financial plans for an organization. These policies and plans are translated into the language of *accounting:* cost, revenue, assets, and liabilities. The information obtained through accounting is summarized for management in the form of profit and loss statements, and, finally, balance sheets.

Both accounting and finance are concerned with optimization of profit—cost reduction and elimination of waste. Therefore they readily lend themselves to a management science approach.

A. Linear and Integer Programming

Linear and integer programming have been used in accounting and finance for a number of years. These techniques are used in investment planning, capital budgeting, and many other areas. For example, suppose a company is considering building a new plant to produce certain products and that the selling price and production and marketing costs are known. The company is required to operate under certain financial, economic, and physical constraints. It may be required to maintain a certain amount of working capital, say, or the market for some of the products may be limited. Such problems, once the profit or cost function and the constraints are expressed as linear functions, could be solved using linear or integer linear programming to show the most profitable way (least-cost method) of operating the plant.

One of the newer applications of linear programming pertains to the assignment of audit staff personnel. Linear programming is used to find the appointment of auditors to check assignments so that the limitations of the audit office under consideration are satisfied while the professional and economic objectives of that office are met.

A linear programming approach will provide needed information for decisions about the following:

1. Scheduling of professional development and educational periods
2. Which staff members should be requested to work overtime
3. Determining a policy for fair and reasonable staff compensation
4. Which clients should be approached for additional work
5. Whether or not to add clients when office staff is working at near capacity
6. How compensation rates or individual auditor activity will affect assignment schedules
7. How vulnerable the office is to loss due to an error in assignments

The only major problem revolves around the benefit-maximizing nature of linear programming. The audit staff must maximize both monetary and nonmonetary benefits. The problem then arises as to how the nonmonetary benefits should be quantified [1].

B. Sensitivity Analysis

Sensitivity Analysis consists of examining the effects of changes in the parameters of a problem. Usually there will be some parameters that can be assigned some reasonable value without changing the optimality of the solution. However, there may also be parameters with likely values that would provide a new optimal solution. This is especially important if the original solution would then have a substantially inferior value for the objective function. The purpose of this procedure is to identify these sensitive parameters.

Sensitivity analysis is usually applied to linear programming problems in one of two ways. The first is to examine the effect of a discrete change in one or more variables. A new solution is then found and the process is repeated until all the feasible changes are exhausted. The second is known as *parametric programming* and requires testing the solution for optimality over a continuous range of value of one or more variables. This method is usually used to identify the upper and lower bounds of the values of a variable for which a particular solution is optimal.

Sensitivity analysis is used in accounting and finance to show how project profitability may be affected by variations in an element of project revenue, operating cost, or investment. A thorough sensitivity analysis will display in graphical or tabular form how much project income and return on investment will differ from the project estimate if sales, labor costs, or some other factor does not turn out as expected.

C. Zero–One Programming

This technique will solve linear programming problems when the variables take the value of zero or one only (see Chapter 6, Section II.D.1.a). This technique is most useful in the capital budgeting problem, where several projects are competing for limited capital.

The objective function is to maximize the net present value (NPV) of the projects under study, with the consideration of budget, labor, material, or any other constraints applicable to the problem. A zero–one capital budgeting model can be expressed as

$$\max \sum_{i=1}^{n} (NVP)_i X_i$$

$$\text{subject to } \sum_{i=1}^{n} P_i X_i \leq B_t, \qquad t = 1, \ldots, T$$

$$\sum_{i=1}^{n} m_i X_i \leq M_t, \qquad t = 1, \ldots, T$$

$$X_i = 0, 1$$

where

n = number of projects under consideration

$(NVP)_i$ = net present value of project i

P_i = initial investment of project i

B_t = budget available at period t

m_i = material or personnel required for project i

M_t = material or personnel available at period t

Interdependencies between projects, such as the following, can be incorporated in the programming model by additional sets of constraints:

Mutually exclusive projects: A set of projects is considered to be mutually exclusive when acceptance of one project from the set makes the remaining projects clearly unacceptable.

Contingent projects: When the acceptance of a project depends upon the acceptance of other projects, the entire set constitutes a set of contingent projects. A set of contingent projects can be combined into a single compound project.

D. PERT

PERT (Chapter 6, Section III.B.4) has been used extensively in finance and accounting. Accountants have been using PERT for a number of years to reduce delays and conflicts within the firm. One area of application has been the estimation of cost and time to complete audits.

Recently, the use of PERT has been advocated in the planning and preparation of consolidated financial statements. This work is usually done under a strict time schedule and under considerable pressure because these statements must be completed and published as early as possible after the close of a company's accounting period. For this reason, cost is of minor importance. However, the larger the number of subsidiaries and minority stockholders a company has, the longer it takes to provide the statements; thus time is the most important factor to be considered. The distribution of work among staff and supervision of the assembly of the statements become very important. The application of PERT to these activities provides an easier way of planning.

E. Break-Even Analysis

Break-even analysis (Chapter 6, Section II.B.1.d) is an analytical approach to profit planning based on the relationships between costs and revenues. The break-even point occurs where sales will exactly cover total costs: No profit is made and no loss incurred.

The first step in the preparation of a break-even analysis is to separate costs into two categories: fixed and variable. *Fixed costs* are those that remain reasonably stable regardless of the level of output, whereas *variable costs* change with the output level.

To define break-even analysis mathematically, certain variables will first be defined:

$$TR = \text{total revenue}$$
$$TC = \text{total costs}$$
$$TF = \text{total fixed costs}$$
$$P = \text{sales price per unit of output}$$
$$V = \text{variable cost per unit of output}$$
$$Q = \text{volume of output in units}$$

With this information, we have

$$TR = P \cdot Q \tag{7-1}$$
$$TC = V \cdot Q + TF \tag{7-2}$$

To find the break-even point, simply equate TC and TR (in equivalent units) and solve for the desired figure:

$$\text{Units at break-even point} = TF/(P - V)$$
$$\text{Costs and revenue at break-even point} = TF/(1 - V/P)$$

The break-even point determined mathematically above may also be determined by *graphical analysis*. The graphical approach shows very clearly the relationship between profit and changes in sales price or changes in fixed and variable costs. Basically, management determines fixed and variable costs and these are then compared at different volumes of activity with the revenue that would be earned at each level, as shown in Figure 1.

Although break-even analysis as shown in Figure 1 gives the appearance of being an exact tool, results should be considered approximate. The reason for this is the assumption of linear relationships among variables. There are many cases where it is reasonable to expect increased sales if prices are reduced, or the reduction of average variable costs per unit over some range of output. These assumptions are best shown in a nonlinear break-even analysis as depicted in Figure 2. Here there are two loss regions, two break-even points, and an optimal point for units of output where profit is at maximum.

F. Simulation

Sometimes mathematical representation of a system is too involved to be solved or the system is so complex that it cannot be adequately described and analyzed by a mathematical model. In these situations, simulation

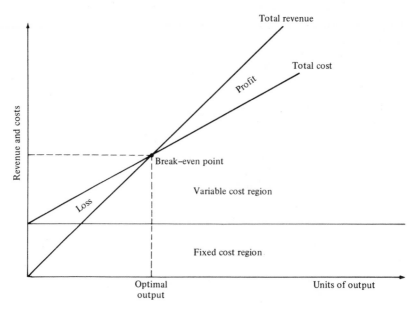

FIGURE 1. Linear break-even chart.

FIGURE 2. Nonlinear break-even chart.

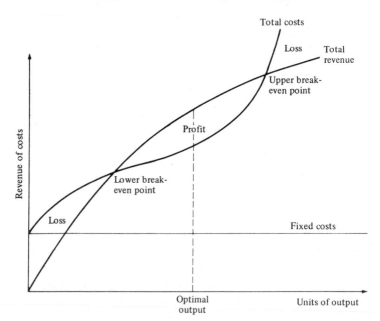

(Chapter 6, Section III.B.8) lends itself as a valuable tool for obtaining a solution.

Accounting and finance have begun to use simulation in a number of ways (as have most other disciplines). Some of the situations in which simulation is used are computing taxes, depreciation, and calculating long- and short-term debt figures. It is also useful in cost analysis and control, capital budgeting, and making replacement decisions. Simulation can even be used in training auditors by producing financial statements of a simulated firm via computer in accordance with generally accepted auditing standards. Thus, simulation is not only being used as a tool to answer difficult mathematical questions, but also as a teaching device.

II. BANKING

The banking industry has trailed other industries in the development and use of management science models for a combination of different reasons, many of which no longer prevail:

1. In the late 1940s and early 1950s, bank deposits were growing and bank profits were satisfactory. However, tighter money market conditions prevail nowadays and prime rates are soaring. This makes it necessary for banks to seek more effective measures and innovative techniques.
2. Management science research requires sizable investments in trained staff and in computer time, as well as long lead time before results can show a payback for the investments. Smaller banks therefore cannot afford outlaying such investments. However, "correspondent banks" have allowed small banks to benefit from their management science developments. In this respect, major banks act as quasicentral management science consultants to small banks.
3. Banks used to lack well-trained staff in mathematical approaches to problem solving, but have now developed such staff through hiring recent graduates and/or training.
4. Banks employed full-scale, up-to-date computers mainly in performing clerical operations of data processing. Significant time is now being allowed for research-oriented problems.
5. Although there have been successful applications of management science techniques in banking, the literature has not reflected much of this progress. This may be attributed to competitive reasons that necessitate withholding publications and/or traditional bank attitudes about loyalties of their employees that prevented them from publicizing their work outside of their organizations. These attitudes are gradually being relaxed and hopefully more publications will appear.

Progress in areas allied to management science, such as marketing research and econometrics, has fostered successful applications of management science in banking. However, there are still some obstacles to banking applications:

1. Management science deals mostly with quantitative parameters; many decisions in banking depend on judgmental qualitative factors.
2. Optimization deals with the choice between alternatives, some of which may have to be identified by sources external to the expertise of the management science staff. A team of specialists from both the banking operations and the management science staff should be formed to uncover these alternatives.
3. Many banking problems become too complex when criteria for optimization are considered. An example would be trying to maximize yield and minimize risk at the same time. Conflicts among alternative criteria may be resolved by involving management either to occasionally sacrifice one objective for the other or develop acceptable ranges for one while optimizing another. Goal programming is a technique used in situations like this.
4. Implications of a management science study will frequently extend far beyond the original scope of a small problem under study. A total systems approach (from the viewpoint of the bank as a whole, rather than from the narrower viewpoint of a single department) can represent difficulty to a management science study.
5. Hostility and resistance to a management science study hinders successful use of its results. However, management science research is now more accepted at both management and operating levels.

In addition to the usual long-/short-term and deterministic/stochastic distinctions, management science applications in commercial banks can be classified according to several major banking functions:

General management

Lending and credit (including deposits)

Bonds (government and municipal)

Trusts (corporate, personal, and pension)

Operations (e.g., check clearance, computer service, truck routes)

A. General Management

Management science techniques can provide much assistance in the general management area of banking. Asset/liability management, mergers and acquisition, demand deposit, management information and control systems, and management games are some examples where management science tools can be applied.

1. Asset/Liability Management

The profitability of a bank depends heavily upon the quality of its asset management decisions. Problems such as the complex nature of available yield/liquidity trade-offs, the multiple and often subtle interactions among balance sheet accounts, and the effect of inherently uncertain future expectations on present actions add to the complexity of asset/liability decisions.

Some analytical management science approaches have been tried by researchers. Their efforts were directed both toward total asset/liability management models and models for subproblems that when combined may form an integral part of a total model. Such efforts include the development of econometric forecasting models for such factors as future deposit levels, loan demand, and interest rates, as well as models to determine optimal investment portfolios, given prior forecasts of these factors. Further efforts by major banks are still being directed toward developing analytical models for the asset/liability management problem. Most banks, however, are still applying intuitive decisions in that respect.

Perhaps before directing all attention toward developing an integral solution to the problem, research should be directed toward solving related subprograms that would pave the way to a total asset/liability planning model. After the development of all necessary relationships, simulation approaches can be beneficial, especially in the first exploration stages. The concepts of industrial dynamics (defined in Chapter 8, Section II.I) are also recommended to represent the dynamic nature of the problem and the many feedback loops that should be incorporated in an asset/liability model.

2. Mergers and Acquisitions Analysis

Major banks make decisions regarding mergers and acquisitions of smaller banks and/or special types of business companies. These banks have developed systematic management science models to analyze and evaluate a prospective merger or acquisition. Such models would analyze future cash flows originating from the institution to be acquired and then determine a justifiable price range and/or a rate of return to equity shareholders, contingent on the price asked by the bank or company in question.

Using these models on conventional computers, to which many banks have access, executives are able to respond quickly and bid and negotiate the acquisition price correctly.

3. Demand Deposit Models

Corporate treasurers in banks and corporations pay close attention to demand deposits of their corporations. Cash management techniques used by corporations have a direct impact on banks' deposit levels. Accordingly, this is an area where management science models can provide profitable solutions. By optimally managing demand deposits, banks and corporations can get more benefit from investing in profitable opportunities of asset investment in interest-bearing securities.

Although not much research has been made public, management science techniques have been widely applied in this area. Linear programming, dynamic programming, statistical analysis, and simulation are techniques that are commonly used for such models.

4. Management Information and Control Systems

Management information and control systems in banking are directed toward gathering and coordinating necessary information and providing management reports. Through these systems, management science techniques can determine the following:

1. *The profitability of different operating units:* This can be decided by evaluating marginal effects and by sensitivity analysis of the different parameters affecting the profitability of these operating units.
2. *The profitability and quality of different banking services:* Management science techniques can be utilized to determine the optimal mix and dimensions of various possible banking services. Analyses of new innovative services in terms of their profitability and effectiveness can be studied and evaluated.
3. *The relative position of the bank in the industry:* This can be best decided by evaluating different planned growth strategies and their feasibility.
4. *Serious and critical situations:* Identification of these serves to guide developing corrective measures. For example, when operating losses of a bank increased significantly, an analysis of their cause as well as minimization studies, can be made.

5. Bank Management Games

There are some bank management games for training future bank executives, senior management, and middle management. These management games are business simulation models that provide a challenging decision-making exercise. Although the use of a computer is required for these games, it is not necessary for the participants to have any knowledge of the computer.

B. Lending and Credit

There are many management science applications to be employed in optimizing the different banking service operations. Mathematical programming, statistical analysis, simulation, and so on have been and still can be applied to maximizing profits, minimizing cost and losses, or reducing risk. The following are problems for which management science has provided successful models and solutions.

1. Credit Analysis and Evaluation

Evaluation of the credit worthiness of individuals or corporations is an important function in banking, due to the risk involved in lending money to unstable business corporations or financially irresponsible individuals. There is an abundance of data available in this field to stimulate management science research. Statistical analysis has been widely used to develop numerical evaluation and numerical credit scoring:

Meyers and Forgy [2], for example, have applied discriminant analysis and regression analysis to determine and assign relative weights to the different questions on a long application form. Smith [3] applied statistical analysis for measuring risk on individual accounts in order to measure and control the quality of credit portfolio and to estimate loss rates. Mehta [4] applied the techniques of sequential statistical decision processes to establish an analytically based control system. Liebman [5] developed a Markov decision model for selecting optimal credit control policies, which he formulates as a dynamic and linear programming problem. There has also been a great deal of unpublished research in this area, much of which appeared to be yielding promising credit criteria. Such criteria, when developed, however, tend to be attacked by both the public and government, whether as discriminatory or for other reasons. Bankers therefore sometimes decline to base their judgments solely on these models. Even when they do use them as a guide, they would not publicize or admit their use in order to avoid criticism.

In the corporate area, models have been developed to analyze and evaluate the credit worthiness and soundness of smaller banks, foreign banks, industrial corporations, and proposed ventures. Hester [6] developed a regression analysis model to investigate the extent to which alternative characteristics of bank loans to business firms (such as interest rate, maturity, etc.) are affected by both the borrower and the bank. Many other econometric and management science models have been developed and are in use by major banks.

Management science models can also be successful in evaluating the ability of borrowers to expand their business and accordingly pay back a possible defaulted loan. These models recommend payment schedules that would enable the borrower to stay in business while paying back the loan.

Management science models can help determine the optimal time to pursue defaulted loans. Applied statistical techniques can be used to indicate how long the adjusters or collectors should pursue these nonpaying loans. Many of the major banks and collection agencies employ some management science models for this problem.

Management science models have been developed (although few have been published) to evaluate loan interest rates for different risks, payment schedules, and future market expectations. Such models have proved successful, especially when bidding interest rates on sizeable loans.

2. International Lending and Foreign Exchange

Most banks, especially major banks, have expanded their business by offering loans to international business, as well as by opening branches in foreign countries. Such expansion gives rise to two major problems:

Foreign exchange exposure from dealing in foreign currencies

Evaluation of international loans and the net earnings which are subject to foreign tax withholding.

In the foreign exchange exposure problem, management science models have been successful in optimizing the portfolios of foreign investments and holdings in foreign currency in order to minimize the effect of currency value fluctuations and possible devaluations. In this area, statistical analysis, forecasting, and stochastic programming can be most helpful.

Loans made in foreign countries are still evaluated by many banks on the same basis as loans made in the national market. Foreign loans, however, are in many cases subject to foreign tax withholding. The receipts of the foreign tax withheld in foreign countries are accumulated and can be used, to a certain limit, to offset the U.S. income tax. Mathematical programming has been employed to provide common profitability measures for foreign loans made in various countries to ensure optimal investment in the loan portfolio.

3. Marketing

Bank marketing departments can apply management science techniques in the following areas:

Branch site evaluation: Statistical analysis, simulation, and mathematical programming can help evaluate the site of a present or proposed branch.

Safe deposit box use: Statistical analysis can be employed to determine the underlying factors related to the marketability of safe deposit boxes. Such studies can highlight factors to be exploited in marketing promotion campaigns.

Evaluation of new services: Management science techniques can be employed in evaluating the chances of success for proposed new services. Also, when a new service is introduced, data about its progress can be fed into a management science model, say, a simulation model, in order to extrapolate and forecast the expansion of the service.

4. Customer Service

As part of their services to their clients, banks have developed and employed different management science models to solve some of their clients' problems. As a side benefit from these services, banks generate business for themselves. For example, acting as a collector for service companies such as telephone and gas companies and publishers, banks can utilize the deposits of their collections. In order to attract such businesses, banks offer various services in return:

Coordination and selection of mergers and acquisitions: In this respect, banks offer data bases on possible offerings for a given merger or acquisition. They also try to help in the pricing of the new acquisition and in drafting the form of the merger.

Determination of the optimal network of local box locations: Mathematical programming, together with statistical analysis, have provided solutions for determining optimal networks of local boxes to expedite the collection and clearance of received checks. These networks take into consideration the location of the company, its customers, and federal reserve banks where checks can be cleared, as well as the effect of associated variables such as mail time and clearing time along the network.

C. Bonds

The asset portfolio of any bank includes bond issues. Some of the commercial banks are also underwriters or dealers in government and/or municipal bonds. Although there is much research that has been developed in this area, little of it has been published. Management science models in this area are designed for the following:

1. Bond issue pricing (the market of bond trading is, in general, very slim)
2. Bond portfolio pricing

3. Bond swabbing and arbitrage
4. Determination of optimal portfolios
5. Determination of optimal coupon schedules
6. Determination of risk premiums

D. Trusts

A lot of management science research has been directed to the various aspects of the investment and management of funds, which include:

1. The distribution of investment portfolios among alternative investment opportunities in common stocks, preferred stocks, corporate bonds, mortgages, and so on
2. The optimal composition of portfolio for maximization of return and minimization of risk
3. Performance measurement of portfolios and their managers
4. Establishing market trends
5. Establishing trends on securities as well as possible risks

Much research is being directed to exploit the stochastic and random nature of the stock market. Each study is trying to find the "golden" criterion which, if used as a rule for stock trading, would result in "beating" the market. On the other hand, none has yet been able to disprove the "random walk" character of the stock market.

Studies in this area have employed a wide range of management science techniques. They involve econometrics, statistical analysis, simulation, utility theory, forecasting, and mathematical programming. The most famous of them is the Markowitz model [7], which tries to maximize the discounted value of expected future returns from a portfolio by utilizing the concept of efficient frontiers over given risk assumptions. Markowitz's work was received enthusiastically by mathematicians for presenting a good mathematical framework and has been extended and modified to apply to other types of problem dealing with expected return and risk. Practitioners, on the other hand, have been skeptical about the practical application of the model. Their major criticism is based on the hypotheses that if all the input data to a Markowitz model could be known or easily evaluated, they would be able to make good decisions about their portfolio without needing to implement any model.

E. Operations

The operations function in banks has a lot in common with those of other industries in terms of nature and content. Management science has been (and can further be) used to solve operational problems such as the following:

1. *Optimization of bank teller window operations:* There are some studies that applied queuing theory and simulation techniques to find the optimal number and types of tellers at various times to reduce teller costs while maintaining or improving the service level. NABAC (National Association for Bank Audit, Control, and Operations) Research Institute has developed a Monte Carlo simulation model [8] as a solution to this problem. There are other models that have been tried and still can be developed as a solution to this problem.
2. *Optimal processing of check clearance operations:* Banks are not able to utilize the funds represented on received checks for investment opportunities until they are cleared through federal reserve banks. The daily received checks for some banks amount to millions of dollars. By speeding the processing of these checks for clearance through the federal reserve banks, banks would have earlier availability of the money and could thus be able to use it for investment opportunities. Management science models have been developed to give the optimal sorting pattern, optimal work schedules, optimal staff size, and so on to minimize processing time.
3. *Optimal routes and schedules for check-carrying messengers:* This problem is important for two major reasons. First is the reduction of the operating costs of these messengers. Second, and probably more important, is its effect on speeding check clearing operations. Mathematical programming, simulation, and/or statistical analysis can be used to solve this problem.

Management science models can be applied to many other operational problems in banking such as optimal scheduling of staff for seasonal work loads, optimal computer configuration for bank processing operations, simulation of services to determine optimal designs and critical paths, and studying the nature of costs underlying bank operations to determine critical areas for improvement.

III. PERSONNEL

Since human beings are individually different, many personnel problems must be handled on a case-to-case basis, taking due account of the particular employee's record, concerns, personality, and so on. Management science methods are not highly effective at this "micro" end of the personnel scale, but can be effectively applied in the forecasting of personnel needs, employee scheduling, payroll and fringe benefit problems, opinion polling, training development, and facilities and policy development. For example, a scientific method may be used:

1. To maximize employee benefits and minimize cost by controlling not only the company's payroll finances but also the employee association funds and financial aids
2. To develop sound information on employee opinion so that intelligent management decisions could be made
3. To determine the best extracurricular education programs to offer to employees
4. To plan sizes of parking lots, routes into cafeterias, morale development programs, and motivation programs.

The one obvious point is that scientific methods can be used in only the dehumanized aspects of handling personnel. Wherever leadership is possible, the scientific methods had best be applied judiciously and with an eye toward maintaining an effective management structure.

Management science techniques that are applicable to the field of personnel include behavioral modeling, time series analysis, and linear and dynamic programming.

A. Behavioral Modeling

Behavioral modeling is a method of assessing behavioral trends in individuals or groups and is commonly achieved using scales of measurement. Attitudes of individuals toward objects or events is measured by assigning scales or numbers to characteristics of those objects or events. For example, the reaction of people toward a political event could be measured by percentages of people approving or disapproving of that event. Generally, scales of measurement are two dimensional. In situations where more than two dimensions are required, multidimensional scaling (MDS) is used. MDS is a series of computer-based techniques to represent the attitudes or discernments of individuals toward those objects or events in multidimensional space.

One major application of behavioral modeling is employee selection: The behavioral traits of prospective employees can be compared to the traits required for a given position. This comparison is a computer operation in which a broad base of behavioral traits is used for the comparison. Using behavioral modeling in employee selection provides a service for the employee, as well as for the employer.

Behavioral modeling can also be used to determine what employee reaction to management decisions might be prior to their decisions announcement. Also, if the established model is given enough broad-base input about company conditions and work situations, behavioral modeling can provide management with much helpful information about employee relations, hopefully allowing management to spot problems in their early stages. The main problem in behavioral modeling is to design models that are psychologically meaningful and sufficiently mathematically tractable to allow efficient and effective methods of analysis to be developed.

B. Time Series Analysis

Time series analysis utilizes historical data to forecast future quantities. To use this method, two basic requirements must be fulfilled.

1. It must be assumed that the future will not differ radically from the past. The degree to which this assumption is true will determine the accuracy of the forecast. For example, a forecast made in December 1972 of U.S. gasoline sales for 1973 would not have foreseen the Arab oil embargo of 1973–1974, a significant departure from prior conditions.
2. The quantity to be forecast must have past data available for analysis. A sales forecast for a radically new product could not be made using time series analysis because suitable historical data are unavailable.

Imbedded in the historical data are the effects of past technological changes, raw competition, and so on; but to what extent will the future be like the past? Furthermore, to what extent can a long-range forecast safely be used for planning purposes? One can compensate for anticipated differences if enough information is known; however, long-range forecasts are especially susceptible to new and innovative changes that were not included in the historical data.

Past data can be adjusted to smooth out historical aberrations. For example, an extremely cold winter may have restricted the sale of a product for a single season. Adjusting the data to a normal or average base will remove the effects of abnormal weather on sales. Similarly, historical data may be used to eliminate or compensate for the effects of other parameters. The result of such compensation is to neutralize the effect of these parameters on the quantity being studied. Known or estimated values of these parameters for the forecast period may then be injected to arrive at the most realistic forecast possible.

Employee forecasts by department, division, or company are typical applications of time series analysis. Historical employee data are generally available, and may be tailored to suit the particular needs of the study. For example, the number of employees may increase temporarily each December in order to handle the holiday sales period. The raw data may be adjusted or normalized to compensate for the annual December increase. Other adjustments may be necessary for other factors (seasonal, weather, cyclic, etc.).

Time series analysis may also be used to forecast such quantities as wage and salary compensation, fringe benefit compensation, and employee productivity. Practical application will depend on the particular company, industry, and proper need of the necessary requirement.

C. Linear and Dynamic Programming

Linear and dynamic programming may be used in almost any personnel situation where an optimal solution must be selected from among numerous alternatives. For example, consider a large corporation with a sizable number of production workers. General practice is to promote a production worker with experience and leadership abilities to the position of production or line supervisor. The production supervisor is responsible for a variable number of people performing jobs that require various skills. Thus, an optimal solution is needed to determine the best experience mix for such a supervisor; that is, how long and at what jobs should production personnel work before being considered for the job of supervisor, given that each production-level job contributes a specified number of points per unit of time toward candidacy. Some production-level jobs would be required work experience, whereas others might be optional. The points allocated to each job may be determined from the work experience of previous excellent supervisors.

The selection of optimal leadership abilities may also be solved by linear and/or dynamic programming. However, since the relationships among leadership variables would be subjectively derived, it is possible that individuals with personality characteristics similar to those of the modeler would become supervisory candidates. Therefore, it would seem more appropriate to address such problems through the selection of optimal work experience.

Linear and dynamic programming may also be used to solve personnel allotment problems that entail meeting definite restrictive criteria such as cost and availability of resources.

IV. PRODUCTION

Production can be defined as the conversion of resources into products and services to meet human needs. In a production system a multitude of interacting parts and considerations exists (see Figure 3). By analyzing each part of the system we can discover the nature of the whole system.

Human performance is probably the most complex and unpredictable factor encountered in a production system. A person's basic abilities are often collected under the broad headings of data sensing, data processing, and data transmission. Equivalently, work can be classified under the three functions of obtaining information, making decisions based on the information, and acting upon the decisions. In order to utilize the human

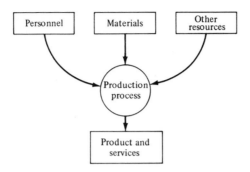

FIGURE 3. Parts of a production system.

factor of a production system effectively, a satisfactory arrangement of production personnel to jobs must be realized. One important concept that has been developed to assist management in job assignment is the job allocation model.

A. Job Allocation

The allocation of individuals to particular jobs is a matter of much concern. Different jobs usually require special skills, attitudes, or adjustment abilities.

One approach to person–job matching is the *assignment method*, a variant of linear programming. This technique is employed when a number of operations are to be assigned to an equal number of operators with the restriction that each operator performs only one operation. Workers might be matched to machines according to the number of pieces produced per hour by each individual on each machine, or by virtue of the expected cost for each person to accomplish the job or project.

In the assignment method the format must be a square matrix, where the elements of the matrix are the costs, profits, or relative ratings for each candidate assigned to a position. Several computer programs are available for use with this technique.

B. Plant Location

The problem of plant location exists whenever a plant is to be built or expanded. In case of expansion there are three alternatives open to the industrial concern: do not expand, expand at present location, and expand at a different location. The nature of the alternative locations and the attractions and disadvantages of each are the determining factors in the final decision.

The primary factors in determining plant location are raw materials, energy resources, capital, labor, transportation, and market. When selecting a site for the location of a manufacturing concern, these primary factors must be present in some combination. In essence, the problem of locating an industry becomes a question of finding a site where the favorable factors outweigh unfavorable factors.

1. Raw Materials

Usually the availability of raw materials is a major locational factor. Raw materials are not evenly distributed; they are localized. The availability of local leather supply concentrated the shoe industry in Massachusetts, and Cincinnati became an early soap center primarily because fats were available from local stockyards. However, in the 20th century the influence of raw materials as a location factor has declined somewhat due to several factors. First, the transportation network has improved. Second, as manufacturing becomes more complex, individual manufacturers process their basic raw materials less than before. Third, technological improvements have lessened the need for locating near a particular raw material. Finally, the general decentralization of manufacturing, combined with competitive sales equalization practices, has encouraged manufacturers to increase the distance between the producing plant and the raw materials source.

2. Energy Resources

Energy has been a major factor in site location throughout history. In the early stages of the industrial revolution, manufacturing regions developed where water power was available. After the development of alternative power sources (e.g., coal), manufacturing was only limited to power sites to the degree that transportation of energy also cost money. In many heavy industries, fuel, when considered a major cost, becomes an important consideration in locating plants. A manufacturing process that requires a large quantity of energy is therefore likely to be located near an energy source. However, there has been some decline in the relative influence of fuel as a location factor. For example, the glass industry is finding it cheaper to transport natural gas to the plant than to transport the fragile glass to market places.

3. Capital

Regional capital availability was a major factor in the past; today, this is true only in some very isolated places. Availability of capital only applies to the early stages of industrial development and only when capital does not enter the area from outside. The growth of corporate finance has greatly reduced the importance of local funds; however, the money available in an area still depends upon the area's earning power. Also, money is more mobile within a nation than when it must cross international boundaries. A major difference between advanced and underdeveloped countries is the existence of highly developed monetary and credit systems, which make it possible to acquire the capital needed to establish an industry. In an underdeveloped nation one must accumulate capital through savings or outside sources.

4. Labor

The cost, availability, stability, and productivity of labor are all important factors in every manufacturing enterprise. The growth and future of an industry is in jeopardy if labor conditions are not favorable. Labor requirements vary from industry to industry. One may need a large number of skilled workers, another a large number of unskilled workers, still another may need a mixture of each.

The cost of labor varies from region to region, with differences between large cities and smaller communities and even between neighboring cities. In general, the lowest wages in the United States are found in the southeast and the highest in the Pacific coastal cities. The wage scale is affected by many factors including the skill of the worker, the types of industry within an area, and competitive demand for workers. In recent years it has sometimes been felt that differentials in wages are disappearing because of such factors as federal legislation and union efforts. However, the geographic spread in wages still continues. One basic reason for this is the expenditure standards of individuals in different areas. In larger cities there is a wider choice of how dollars may be spent, which results in an increased inducement to spend.

When labor costs are an important portion of the total costs of a product, the labor cost of the area may be very critical. There are two types of location that can be considered as low labor cost locations. One is where population has expanded faster than employment opportunity and therefore people will be willing to work for less than elsewhere. The other is where the local labor force is highly productive, resulting in a higher productivity per person-hour.

The availability and stability of labor are very important factors when locating. If a large supply of unskilled labor is required, it may be best obtained in a metropolitan area. Generally, however, unskilled labor is more mobile than skilled labor (rarely tied down, willing to move to a metropolitan area). Labor stability can generally be measured by its past history. An area with a history of industrial peace is much better than one with a history of long labor strikes. The attitudes of the employees, employers, and unions can go a long way in determining the degree of labor stability.

5. Transportation

The cost and availability of transportation greatly affects the location of a manufacturer. Transportation facilities bring regions together, and the location of an industry in a given area may depend directly on the type of transportation if freight rates are a major portion of production cost.

In the early stages of industrialization, when plants received raw materials locally and finished goods were marketed in a more restricted area, transportation played a lesser role in locating plants. Manufacturing then was highly decentralized. As transportation facilities increased and became more efficient, the delivery of raw material and distribution of finished products outside the local market became an important consideration. Therefore, in selecting a manufacturing site, the availability of adequate and economical transportation facilities is a major factor to consider.

6. Market

The market is increasing in importance as a location factor. If there is a substantial local market, there is frequently a cost advantage. If there will be a large weight increase in the product by the use of some material that is available in most places (such as water), locate as near the market as possible to reduce distribution costs. The same is true if the bulk of the product is increased in the manufacturing process. If the item is fragile or perishable, a market-oriented location is desirable. Also, locating near the market may be desirable if the product is of low value: Transfer costs could become a major portion of the selling price.

7. Secondary Factors

There are also secondary factors to be considered when locating an industry. The influence of the physical environment on locating an industry has changed with time. In the early days of the aircraft industry, level land was sought; today it may not be a factor. Still, some manufacturing does require a certain lay of the land. For example, shipbuilding requires a gently sloping oceanfront setting.

The importance of climate as a factor depends on the industry. The warmer regions of the United States, however, have a comparative cost advantage over colder areas. For example, a southern industrial worker might have a salary 10%–20% lower than a worker in New England, but still have the same standard of living because heavy winter clothing is not needed, heating bills are smaller, housing costs are lower, and some food can be obtained locally for less. How the local atmospheric conditions handle pollution is an important climatic consideration. If it stays or hangs in the area without being dispersed it can be a very serious health problem. Also, adequate and proper water supply (precipitation levels) should definitely be considered when locating.

Local, state, or national governmental policies may also influence location and growth of all types of industry. At the national level, the government frequently attempts to improve economic conditions in some areas through legislation and legal actions.

The availability of managerial skills and personal considerations are other factors that enter into the final location decision.

C. Plant Layout

Plant layout is a companion problem to plant location. There are three basic types of layout: product, process, and fixed position.

Product layout deals with the characteristic of mass or continuous production. Here a logical sequence of operations is used to reduce material handling and inventories, lower the production cost per unit, and make it easier to control and supervise the product that moves along an assembly line.

Process layout involves a functional arrangement of job and batch production. This allows good flexibility and reduces the investment in machine and equipment, but increases handling, space requirements, production time, and the need for closer supervision and planning.

Fixed-position layout involves facilities centered around a static product. Workers and machines are brought to the product, as in the construction of dams, bridges, and buildings.

Several computer programs utilizing mathematical modeling techniques, such as the mixed integer algorithm, are available in the area of plant layout.

D. Management Science Techniques in Production

Analytical techniques, as well as statistical and simulation models, can be used in production systems to improve and facilitate the performance of the job in every part of the system. In this section, a few of the numerous techniques applicable to a production system are discussed.

1. Linear Programming

There are many situations in the production area to which linear programming techniques are applicable, including the following:

Product allocation: In a production shop there may exist a number of jobs that may be run on any of a number of machines. By using linear programming, it is possible to determine how best to allocate work to machines so that the total run time or the total production cost is minimized.

Product mix: A production facility may be able to produce several different products, each of which has different costs associated with its production. Each product will also have a different market demand and selling price. With this information and the knowledge of availability of resources, one can determine what quantities to produce of each product so as to maximize profit or minimize cost.

Blending: Linear programming is used to determine what ingredients should be used and in what proportions in order to meet the demands for specific products (e.g., oil, gasoline, alloys).

Distribution and shipping: Given a product demand at various locations, a supply of products at several warehouses or production facilities, and the costs associated with product delivery from each source to each destination, it is possible to apply linear programming to determine which source should ship to what destination and in what quantities so as to minimize the total distribution cost. (Recall the transportation model of Section II.D.1.a in Chapter 6.) Alternatively, given several sources of raw materials and several production facilities with specified demand, one can determine how much should be contracted from each supplier for delivery to each of the production facilities.

Production scheduling: A very useful application of linear programming is related to CPM and PERT network analysis (Chapter 6, Section III.B.4). It is not unusual that a project is delayed so that it becomes infeasible to complete it in the original expected time. If a deadline or due date must be met, it may become necessary to shorten or crash some of the activities in the network. Each activity should have associated with it a maximum amount by which it may be reduced and a cost associated with its reduction. Linear programming can be used to determine which activities should be crashed and by how much so that the completion date is met at a minimum increase in total project cost.

2. Dynamic Programming

Because dynamic programming allows a large problem with many variables to be split into subproblems, each with only a few variables (see Chapter 6, Section II.D.2.b), it is particularly useful for problems that involve a set of interactive decisions made over time. For example, production scheduling decisions made in one month affect the resultant end-of-month inventories, which in turn affect the production requirement for the following month.

If at the beginning of a year we were to consider all production scheduling strategies for each of the 12 ensuing months, the resultant problem would be enormous. The dynamic programming approach allows us to begin by only considering the schedule for the first month, solving that relatively simple problem, then solving the schedule for the second month with the knowledge of the last month's schedule, and so on.

Dynamic programming is applicable to many time-oriented problems, such as periodic maintenance of equipment decisions, inventory replenishment, and equipment replacement.

The practical limitation on the dimension of the state variable is that dynamic programming tends to narrow the focus of the problems solved by the technique. Only infrequently can all the salient factors in an economic problem of any magnitude be boiled down in such a way as to make feasible the numerical solution to a dynamic programming formulation. This realization does not detract from the value of dynamic programming; rather, it puts into focus certain features that distinguish typical applications of dynamic programming from those of linear programming.

Specifically, many real applications of dynamic programming deal with operating decisions: when to order, when to make a machine setup, when to replace a piece of equipment, and so on. Many of these decisions are ordinarily delegated by top management to lower echelons. Therefore, implementation of a dynamic programming model involves either using a computer to assist in routine decision making or providing lower echelons with tabulated rules based on an optimal strategy calculation. The limited impact of a single decision found by the usual dynamic programming applications and the ease with which this decision can be determined are the very reasons for the technique's success.

3. Quality Control

Concern for quality of products and services should exist throughout an organization. Managers at each level of the organization deal with quality in one or more of its aspects: policy establishment, product and service design, design of operating systems, output of the product or service, its delivery to customers, and its service while in use.

In any quality control system, such as the one portrayed in Figure 4, we see that quality begins with managers at the highest level of the organization who determine the basic strategy of the enterprise and its position regarding the level of product and service quality. Policies designed to implement this strategy are based upon long-term objectives, the competence and resources of the firm, market requirements, and technological

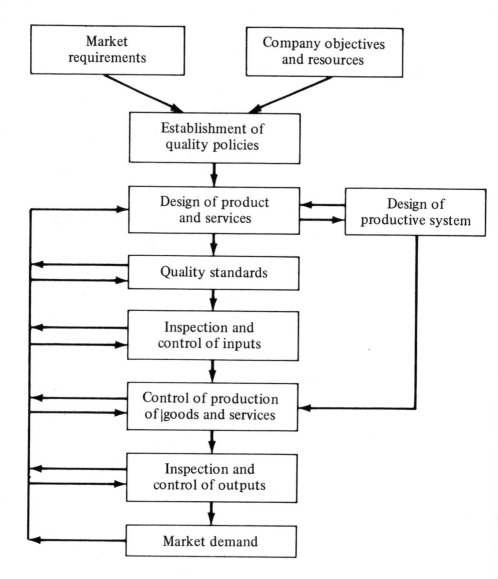

FIGURE 4. The quality control system.

environment. Assessment of these factors will serve to determine what quality levels can be offered and at what price.

In setting quality policies, managers will seek answers to such questions as how do consumers evaluate quality of products and services? what is the technical capability of the firm to provide a particular level of quality? and what is the cost of providing it? In some industries, more careful attention to quality matters is being enforced by government regulation.

The nature of quality is an interesting issue. Consumers view quality in terms of functional and esteem characteristics, and companies design and create products with imperfect knowledge of those criteria. Defining detailed quality standards for some products is very difficult and usually done somewhat arbitrarily. Motivational programs, such as zero defects, can instill an improved quality consciousness on the part of workers.

The measurement of any characteristic or attribute of a set of parts or products will result in variability of the measurements. Statistical quality control procedures set up hypotheses about this variability; when actual data indicate rejection of a hypothesis, the conclusion is that the process creating the characteristic is out of control. Limits on statistical control charts (Chapter 6, Section III.B.6) are typically set at plus and minus three standard deviations ($\pm$ 3 s.d.). Hypotheses for variables are made for both the *mean* (measure of central tendency) and the *range* (measure of variability). Control charts can also be constructed for good–bad assessments or for the number of defects per unit. Construction of actual control charts is facilitated by tables of quality control factors which can be applied to sample data.

The 3-s.d. limits result in a probability of .0027 of rejecting a hypothesis when the hypothesis is indeed correct. This kind of mistake, rejecting a true hypothesis, is called a type 1 error. A type 2 error is committed when a hypothesis should be rejected but is not (accepting a false hypothesis). The *operating characteristic* (OC) *curve* depicts the probabilities of committing type 1 and type 2 errors. Different OC curves are associated with different sampling plans; larger sample sizes discriminate more effectively than do the smaller ones.

Acceptance sampling is concerned with the after-the-fact evaluation of an entire batch of an item, as opposed to controlling the production of that item while it occurs. The underlying methodologies are the same, but the objective is to determine the sample size and decision rules in order to ensure compliance with stated type 1 and type 2 error probabilities.

Some other tools used in quality control are Bayes' theorem (Chapter 3, Section III.A), random number generation, probability distributions, probability theory, statistical tests, statistical estimation, conditional probability, and statistical control models.

4. Queuing Theory

As should be evident from the discussion in Chapter 6, Section III.B.5, there are many opportunities for the application of queuing theory to a production system. Material handling and material flow problems present themselves, as do routing problems, idle times, conveyor lengths, and conveyor speeds.

5. Simulation

A wide range of manufacturing and production problems are suited for simulation techniques. For example, bottlenecks resulting from a change in existing operations may be anticipated. Any proposed change in a system, such as rerouting, altering the scheduling procedures, adding facilities or personnel, and adding a new product, changes the dynamics of the process. Simulation can pinpoint potential trouble spots and, through successive iterations, allow for additional changes so that minimal disruption of operations is encountered when changes are implemented. Simulating the behavior of a proposed plant prior to its construction can aid in avoiding costly mistakes.

Labor allocation has been another area in which companies have benefited from simulation. For example, consider a plan to increase production of an item through use of additional labor. Three alternative courses of action are available: hire more employees, work overtime, or work with second-shift operations. Comparing the cost of each alternative with the increased output indicated can be done by simulation in order to reach the most profitable decision. (See Chapter 3, Section III.B on decision tree analysis.)

E. Measures of Production Operations

A proper way of reviewing how efficient an operation is performed is by subdividing the task into basic components. There are several methods available to analyze the breakdown of the operation, including process analysis, the principle of motion economy, and sampling theory.

1. Process Analysis

The objective of *process analysis* is to improve the sequence or content of operations required to complete a task. This method makes use of three different types of techniques:

Survey charts help in categorizing present procedures in the initial phase of an investigation. Flow process charts, link charts, and organizational charts are examples of survey charts.

Design charts are used to develop changes and to recognize new concepts which in return will improve operations. Human–machine interaction charts and project planning networks belong to this category.

Presentation charts summarize and clarify proposals in order to improve communications. Gantt and time charts are the examples of presentation charts.

2. Principle of Motion Economy

Motion studies attempt to make work performance easier and more productive by improving bodily motions. The *principle of motion economy* pertains to the efficient distribution of work over different parts of the body, preferred types of motion, and the sequence of movements. Some of the objectives of motion economy are as follows:

Minimize the eye fixation time required for perception when time is a controlling factor by using color, shape, and size coding

Take advantage of natural rhythm, arranging work to allow easy, continuous, and repeated movements that develop good work habits

Minimize the number of movements by eliminating unnecessary motions and by combining movements.

Balance the motion pattern by using symmetrical motions in opposite directions and avoiding sharp changes in direction.

3. Sampling Theory

Sampling theory is based on statistical techniques. By using this theory we can analyze work performance and machine utilization by direct observation. This is done by taking a reasonable number of observations of a process at random intervals and categorizing the observations according to the state of the process at the instant it is observed.

Work sampling, which is the field of sampling theory most used, presents several advantages over other techniques. It is used for estimating unavoidable delays to establish delay allowances, for investigating the utilization of high-investment assets, and for estimating the distribution of time spent by workers on different job activities.

The main advantage is that by using work sampling techniques all of the above-mentioned information can be obtained without interrupting the process. The accuracy of a work sampling study depends on the randomness and size of the sample as well as on how well each state of the work is defined.

V. INVENTORY

Inventory is a companion problem of production. Inventory control systems are basically designed for *protection* and *economy*. *Protection* is the provision of sufficient material to meet demands of the particular raw materials, fabricated parts, or finished products with a minimum delay. *Economy* is effecting lower product costs by realizing the savings resulting from longer manufacturing runs and from purchasing larger quantities per order.

Sales forecasts, translated into production and inventory requirements, decidedly affect the cost of inventory. Excessive inventory would usually result in a short cash position, a particularly serious condition for any company with limited working capital. On the other hand, insufficient inventory will increase costs of operation due to work stoppages, increased handling of materials, and small-quantity rush orders.

An important consideration for any inventory control system beyond basic accounting procedures is that of forecasting future inventory requirements.

The importance of good inventory control to an organization is universally recognized. There is no question that better control of inventory may hold the key to problems in financing, marketing, production, and so on. This key position makes it necessary for management to maintain an attitude of perpetual surveillance concerning inventory; change may occur at any time requiring a revision of the methods of controlling inventory if the company is to maintain or better its position. Inventory, therefore, presents a recurring opportunity for improvement. This opportunity can be exploited most effectively by having a good overview of inventory, its objectives, and constraints.

A. Optimal Inventory Policy

The definition of an *inventory system* is essential so as to have a frame of reference and an indication of the scope of the subject matter to be studied. An inventory system consists of the properties that describe the nature and the characteristics of the environment with respect to a particular inventory problem. The properties are essentially assumptions pertaining to the characteristics of the parameters and variables of the inventory environment.

A set of decision rules that minimize the sum of the costs associated with the properties of an inventory system are referred to as *optimal inventory policy*. Optimal decision policies are obtained by the use of inventory models.

An *inventory model* is a mathematical representation of the inventory system's properties and interrelationships. In order to construct a model, the associated properties and resulting interrelationships must be specified or assumed.

There are three phases of inventory analysis:

1. Determination of the properties (specification of assumptions) of an inventory system
2. Mathematical model formulation and manipulation for an optimal solution
3. Analysis of the solution to evaluate what additional costs would be incurred if nonoptimal inventory policies were to be used

All models for inventory decision making are either deterministic or probabilistic, or somewhere in between. Probabilistic models are usually quite complex and use various forms of numerical analysis to arrive at optimal solutions. More common deterministic models assume that demand information is known with certainty and will be constant over time. Not all inventory problems present themselves this simply; however, the deterministic economic lot size models provide a basis for deriving more complicated models.

B. Economic Lot Size Models

These models assume the simplest possible cost structure, involving only order cost and inventory carrying cost, and exclude the occurrence of negative inventory levels. Characteristics and assumptions of this type of model are as follows:

1. Demand D is constant and independent of time
2. As long as the inventory level is positive, the output rate is held equal to the demand rate, resulting in the continuity of the process
3. Lead time L is fixed and independent of quantity
4. Ordering time is to be signaled at a critical inventory level P, the reorder point
5. There is an order cost or setup cost S of input which may depend on the total input quantity, considering quantity discounts and so on
6. An inventory carrying cost C accounts for fixed charges, holding or storage costs, and declined investment cost (lost interest)
7. Input occurs at a constant rate for a time period T determined by the order quantity Q, and the concept of instantaneous input is admitted

Figure 5 portrays the input quantity Q, reorder point P, time between orders T, and the time between reorder and arrival of input, i.e., lead time L. The use of lead time indicates the timely recognition of the fact that when the inventory level reaches the reorder point it is time to reorder.

To determine the optimum inventory policy, an expression for total cost can be derived as a function of quantity for a defined time horizon. The total cost function is made up of two separate cost functions:

Total cost = total order cost + total inventory carrying cost

$$TC = (D/Q)S + \tfrac{1}{2}QC \tag{7-3}$$

where D/Q represents the stock activity in a defined time horizon and $\tfrac{1}{2}Q$ represents the average inventory on hand. The total cost function can be

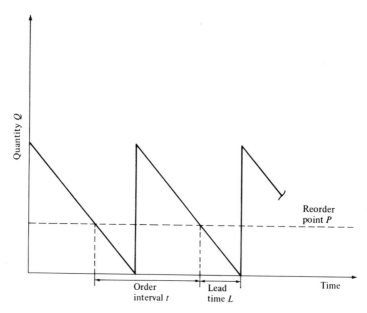

FIGURE 5. Economic lot size model of inventory control system.

minimized with respect to quantity in a number of ways:

1. If each cost function is plotted and then added together as shown in Figure 6, an optimum point (minimum cost) can be decided graphically.

2. Figure 6 illustrates that, when the number of units that are ordered at one time increases, the inventory carrying cost rises, but the order cost decreases. Thus, in this case, the total cost is minimized when the total inventory carrying cost is equal to total setup cost. Therefore, we can say the total cost will be at a minimum when

$$(D/Q)S = \tfrac{1}{2}QC \qquad (7\text{-}4)$$

Solving this equation for Q, we have optimal quantity

$$Q^* = \sqrt{2DS/C} \qquad (7\text{-}5)$$

3. One way of determining Q^* is by trial and error. Various values for Q^* are assumed and substituted into the total cost equation until the minimum cost point is found.

4. A more common method of finding Q^* is by taking the derivative of total cost function with respect to Q and equating this derivative to zero. Thus

$$dQ/dTC = -(DS/Q^2) + \tfrac{1}{2}C = 0 \qquad (7\text{-}6)$$

Solving for the optimal lot size, we have

$$Q^* = \sqrt{2DS/C} \qquad (7\text{-}7)$$

It should be noted that the above equation gives the same results as Eq. (7-5).

In some cases, the assumption that the entire lot is produced or delivered at one time is not true. Often, production or delivery of the total order quantity takes place at a uniform rate over a period of time. This would result in an inventory control pattern as depicted in Figure 7.

The procedure used in this situation is basically the same as that for the previous model except that production or delivery rate is finite and greater than the consumption or demand rate. Therefore, the optimum quantity formula that would result in the minimum cost becomes

$$Q^* = [2DS/C(1-D/R)]^{1/2} \qquad (7\text{-}8)$$

where R represents the production or delivery rate and D is the consumption or demand rate.

FIGURE 6. Graphical model of a simple inventory system's costs.

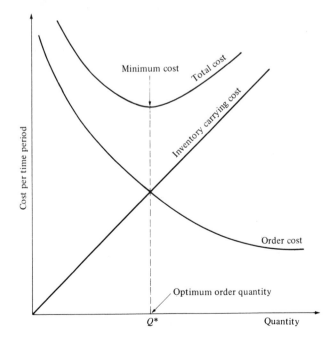

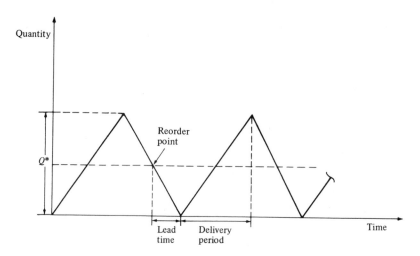

FIGURE 7. An inventory control system when delivery takes place over a period of time.

1. Maximum–Minimum System

An important implication of the economic lot size is the *maximum–minimum system*. When accurate demand information is known, this system combines order quantity (maximum stock level) and order point (minimum stock level) as shown in Figure 8. Stock controllers place orders in such a manner that stock stays within the two specified limits.

In some cases minimum inventory is used to serve as a safety stock for protection against unusual demand or lead times; this system can be described schematically as illustrated in Figure 9.

FIGURE 8. Maximum–minimum system of inventory control.

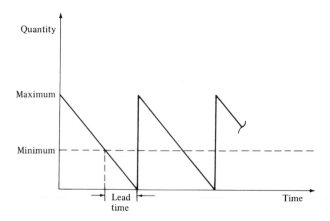

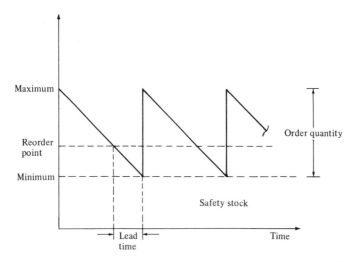

FIGURE 9. Maximum–minimum system of inventory control with safety stock as a protection against unusual demand rates and lead times.

In this situation, the reorder point is calculated with demand rate, lead time, and minimum inventory in mind:

$$P = DL + \text{minimum inventory}$$

It should be noted that when the reorder point equation is used, the same time units should be considered for expressing both demand rate and lead time.

2. ABC Inventory System

In many organizations, usually a small number of inventoried items account for the largest proportion of total inventory value. This fact has led to the establishment of the ABC system, whereby inventory is classified into groups of high-value class A, medium-value class B, and low-value class C items. Usually 10%–15% of inventory items will be class A, about half will be class C, and the balance will be medium-value class B. Effective inventory control can best be maintained by buying the high-value A items in minimum amounts as actually required and buying economic lots of the medium-value B items; these two categories cover about 80%–90% of the inventory value. Class A, in fact, will probably account for as much as 70% of total inventory dollars. Low-value items, even if bought in large quantities, will have little effect on inventory value because they represent a small percentage of the total. Figure 10 portrays the ABC system of inventory control.

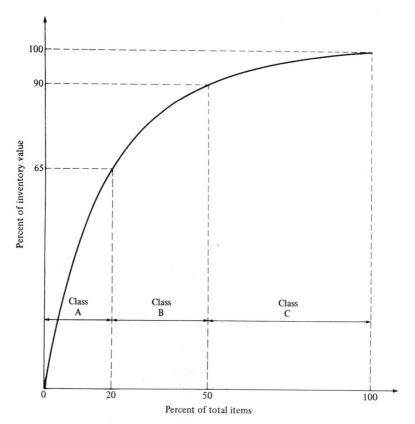

FIGURE 10. The ABC inventory system.

C. Application of Lot Size Models

In applying the economic lot size models to inventory control systems one may experience certain difficulties:

1. Estimation of setup costs, holding costs, lead times, and demand rates is not easily made, and a possibility of error is always present.
2. Calculated optimum lot size may result in an order quantity so large that it is not practically applicable. As a result, optimum order quantity is not used by the production personnel, which distorts the timing and requirements of all components.
3. The model is often applied to all parts regardless of the characteristics of different parts. In general, there are many parts that are not manufactured in lots.
4. Determining a lot size for upper-level components may introduce irregularity in the production of lower-level components. This may create undesirable peaks in the production schedule of these components.

Some general guidelines are helpful in bringing a degree of realism to the application of lot size models. In brief, more than one technique should be used, depending upon the characteristics and nature of the system. For example, when setup cost is important, a variable quantity can be ordered at a fixed number of time intervals. If holding cost is important, a fixed quantity can be ordered at different time intervals.

VI. PURCHASING

Purchasing can be defined as the process of supplying adequate materials, equipment, and services of the right quality essential to the operation of an organization. In general, purchasing includes buying the items needed at the best price possible by screening vendors, handling negotiations required to fulfill the commitments of the company, awarding contracts, and ensuring delivery.

A. Purchasing Objectives

The main objective of purchasing should be supporting the company operations with a continuous flow of materials or services. However, this objective must be attained by keeping inventory investments at a minimum.

Another objective of sound purchasing is the maintenance of adequate standards of quality for items procured. The objective is not to buy goods with the highest absolute quality but those that are best suited for the purpose of the company.

The purchasing department must avoid inventory losses due to duplication, deterioration, waste, and obsolescence with respect to the various items purchased.

In the procurement of capital equipment, the major task is the coordination of all interested parties who can contribute information to aid the buyer in making a correct decision. This is especially important since mistakes in large capital equipment decisions have long-lasting consequences.

B. The Purchasing Function

In general, the purchasing function may be categorized into four areas of activity:

1. Planning, scheduling, and policy making
2. The coordination function
3. Implementing the purchasing function
4. The control function

1. Planning, Scheduling, and Policy Making

Planning is a major aspect of the purchasing function, involving decisions on what to buy, where and when to buy it, and how much of it to buy. In order to make these decisions on a sound basis, one must have adequate data research, reference, and analysis systems. Most items are purchased on the basis of a particular set of specifications. It is the responsibility of the buyer not only to adhere to the specifications, but to make them a part of the purchasing department's permanent file. The specifications are often important in the area of quality control and can function as bases for negotiation in situations where there is disagreement with the source.

The buyer should also maintain complete technical files for each job area in order to have a ready source of reference. The care with which the catalog files and technical files are kept is an indication of the efficiency of both the buyer and the department.

One of the buyer's most important jobs is the location, investigation, and development of new and alternative sources of supply. This activity is part of a continual purchasing research effort that occupies a large portion of the buyer's time. It is a function that should never be regarded lightly because it is the lifeblood of the purchasing effort.

The purchasing department should conduct cost and market analyses. Market studies are important in times of potential product storage, both for the development of new company products, where material procurement could create problems, and in the search for new or alternative materials.

There are some cases where the buyer should conduct *make-or-buy* studies. The make-or-buy decision is usually an important one in the operation of a company. The buyer involved in such a study should be advised to subdivide the information areas and to seek resources within the company, utilizing such specialized areas as accounting, engineering, production planning, and traffic operations. Purchasing should act as a control center for the incoming data; that is, it should compile the reports from all the contributing agencies, add its own report, arrive at a recommendation, and forward it to the proper parties.

Purchasing must establish how much of a material is required before arranging the purchase. Such data can come from a variety of places, depending upon the type of business involved and the type of material needed. For example, for a custom-made item, the estimate should be accurate and generally tied to the individual order. In the case of more standard types of material, the purchase can be programmed to cover the needs of a given time period.

As mentioned earlier, decisions concerning procurement of capital equipment should be carefully planned. The principal decisions in this area regard purchasing new or used capital equipment and whether the equipment should be leased or bought. How these decisions are made depends upon the results of a company's consideration of such fundamental factors as availability, delivery time, costs (both immediate and long-range), and the firm's tax position after allowance has been made for depreciation, cash flow, and the like. Often the ultimate decision is the result of the work of many groups in the company. The buyer would do well to prepare a checklist of the various departments that should be asked to contribute data for the final decision.

The buyer should be aware that certain situations can arise that make the timing of a purchase important. This is usually the case when the item or commodity desired is produced in a fluctuating market, that is, one that rises and falls according to changes in supply and demand. Such markets include many raw materials and commodities for which trading and buying are transacted on the basis of future delivery of an item.

Timing the purchase is an almost automatic reaction of the experienced buyer, but it is often a problem for the new buyer. What is involved is a consideration of several factors: When do we need the item? Are there any raw material problems? What is the procurement lead time? How much time is needed for transit? What is the delivery history of the source? What is the overall effect of the schedule on the purchase of the item? In order to answer these questions, purchasing assumes a consulting role along with production to schedule the activities required in timing the purchase, which is part of the daily workload of the buyer.

Purchasing is responsible for inventory control and must review the requests of the various departments for materials. Such a review generally includes the routine items required to run the business. Often, this problem of checking can be alleviated by establishing the inventory levels in advance and by creating reorder points on the basis of the replacement time schedule. There are usually some guidelines that the buyer can follow: definite estimates for a given job, or a production schedule from which quantities can be worked out; often a sales estimate will be the basis for figuring the quantities to be purchased. Too often, when the quantity guidelines are not definite, the tendency is to purchase the maximum amount to get the best possible price and hope that the future will take care of itself. This is a common error and should be avoided. The usual result is that a large portion of the items may eventually be written off and junked at a greater loss than would have occurred if a lesser quantity had been bought at a higher price.

Scheduling of the various activities involved in a purchase is another important part of the purchasing function. The schedule will include, for example, requisition checking, determining when to place the purchase order, and acquisition of materials needed to make the purchase (e.g., specifications, art work, test results). Included in the purchasing schedule should be the dates for any in-process quality inspections, the final inspection, the delivery date, the traffic data required, and the completion of the receiving process.

An important *policy-making* function of the purchasing department is participation in the preparation of the budget. The key factors involved are a knowledge of the overall company plan and a keen perception of the probable market and price trends in the fields in which the buyer is operating. Most companies operate on the basis of a standard cost system, wherein the budget for the next fiscal year is prepared well in advance. Usually, when the standard costs are computed and the budget is locked in, no changes can be made to compensate for pricing alterations.

2. The Coordination Function

The purchasing department is responsible for coordinating activities both inside and outside the company. The buyer, depending upon the type of business, can expect to be involved in coordinating the activities of vendors, production engineering, packaging development and design, product development, sales, advertising, quality control, and top management. The various types of coordinating may be categorized as follows:

Information exchange: In the vendor–purchaser relationship, the transmission of information is most important for defining specifications, setting up buying policies, establishing quality requirements, negotiating the terms of a contract, and generally establishing and maintaining good source relationships.

Materials standardization: The process of arriving at a standardization of parts or materials is usually the result of a coordinating effort on the part of the buyer. It does not matter who originates the idea; it is the buyer's job to follow through with it. In such a process the buyer will probably have to deal with top management, production, quality control, engineering, marketing, and the vendors. The task of standardization is often a difficult one because of resistance in other areas. In the event of such resistance, the *value analysis* (see Chapter 8, Section II.J) approach can be very helpful.

Inventory control: In this area, as in others, the purchasing department is responsible for coordinating the activities of other departments. For example, plant inventories are usually the responsibility of the plant manager or the production department, and yet purchasing must have the inventory data in order to do its job properly. Getting this information demands a coordinated effort on the part of purchasing. In addition, consolidating requisitions from a group of plants involves not merely a system of procedure but close cooperation among the plants and careful coordination of their activities by the purchasing department.

3. Implementing the Purchasing Function

In general, *implementation* procedures include all the actions the buyer must perform to fulfill the purchasing function. Included are such chores as interviewing salesmen, placing contracts, selecting vendors, analyzing bids, and negotiating for various types of materials and services. Implementation also includes such jobs as scrap disposal, traffic management (if applicable), and any purchasing done for the benefit of the employees.

There are several types of contract involved in purchase implementation:

Materials contracts: These cover a wide area ranging from those for raw materials to those for finished parts and packaging materials. Information can be obtained from technical associations, the company laboratory, or the engineering, production, sales, and legal departments. Some of the most valuable and often overlooked sources for information are the vendors and their technical staffs, who have a real interest in assisting the buyer.

Service contracts and agreements: Depending upon the type of service desired, the department that will benefit should be a party to the making of the arrangement. For example, if computer services are needed, the controller, accounting, and management information system (MIS) group are necessary parties to the preliminary discussions. The key here is to be certain to consult all the possible sources of information.

Subcontracts: Particularly for complicated components, subcontracting is often a large part of the buyer's job in certain types of industry. The major problems seem to arise in the areas of specifications, materials to be used, acceptable alternative materials, testing, quality control, and delivery schedules.

4. The Control Function

Normally, *control* involves some type of checking activity as well as a system developing a history of all purchasing activities. The control phase is extremely important because it is usually the last step in the chain of events that finally culminates in a purchase.

In order to perform the control function efficiently, the purchasing department should develop a filing system. Although files are a major control responsibility, there are no ground rules for what kind of files to keep or how to keep them. There are, however, two major types of file that are necessary: a potential-vendor file and a record-of-purchase file.

A fundamental aspect of the purchasing operation is *expediting*, which is concerned with follow-up problems such as ensuring that the vendor adheres to the agreed delivery schedule, as well as with trying to improve the schedule. Expediting can be as far-ranging as company policy will allow. A good expediter can relieve the buyer from a tremendous burden. Normally, the expediter's position is used as a training ground for future buyers since it is a good place to get the feel of the entire industry. Expediting also offers an opportunity to learn the techniques of dealing with people, the art of persuasion, basic negotiating techniques, and planning and operating techniques.

Invoice handling is another vital part of the final-checking mechanism. Normally, in progressive companies, the accounts payable department handles the actual payment of the invoice. However, the buyer is involved in the process to check prices and to do anything else that the company thinks is necessary during this phase of the operation (e.g., flagging items that are variances, assigning code numbers that identify the reasons for variances).

Vendor relationships also fall under the heading of control activities— specifically, the adjusting and settling of complaints and the evaluation of vendor performance. The settling of complaints is a problem of great complexity and is dependent on the philosophy of the buyer, the relationship between the buyer and the vendor, and the policy of the company in general. The file systems mentioned earlier are, of course, quite useful in vendor evaluation.

It should be noted that the aforementioned control activities are geared to the larger corporation. In smaller companies the buyer may also be involved in other functions, such as operating the storerooms, handling the finished goods inventory, operating the materials cost accounting systems, handling the traffic department as a part of the purchasing department, filing transportation claims, and even physically inspecting and testing incoming materials.

C. Operational Tools and Procedures

Every business must have some records and procedures to keep track of the daily flow of activity; exactly how many there are and of what kind are matters of individual department policy, specific job requirements, and individual preference and necessity. Other factors involved include the type of company, the types of material being purchased, the number of one-time purchases, and the amount of repetitive buying.

1. Purchase Orders

The first and most basic piece of paper is the *purchase order*. Purchase orders generally contain a certain amount of standard information: source name, where to ship the goods, payment terms, and any other items the company deems necessary.

The purchase specifications are the key to the purchasing function. It is here that the buyer must spell out exactly what he is ordering and what kind of order it is. In general, the type of order will determine what kind of instructions should be included in the order form. There are several basic types of orders:

The blanket order: Essentially an incomplete order that really becomes effective only when the vendor receives a shipping release, the blanket order authorizes the manufacture of an item, but contains no shipping instructions. Use of the blanket order is very common in the purchase of packaging supplies, where the buying is centralized and more than one manufacturing unit uses the item. The procedure is for the purchasing department to place a blanket order to cover the requirements of all plants using the items in a given period.

The regular purchase order: This order is used for one-time buys and is generally the same as the blanket order form, although the specific instructions differ somewhat. Normally, the regular purchase order will be used to buy items that are already manufactured or are stock items available for immediate shipment.

The make-and-ship order: Usually, this deals with a specific item that must be manufactured according to a given set of specifications. It is often the result of some form of negotiation, and there may be rather lengthy special instructions.

There has been a substantial effort in the past decade to reduce purchasing costs. As a result of this dedication to the general improvement of the purchasing function, a number of interesting approaches have been developed that are designed to reduce the work load of the buyer. The general areas of major changes are in upgrading and improving the training of personnel and organizational planning. The following are some of the new methods currently being tested: traveling requisitions, stockless purchasing, systems contracting, Dataphone® purchasing, and the cash purchase orders.

Stockless purchasing and systems contracting: In the case of maintenance, repair, and operating (MRO) items, these techniques offer excellent possibilities for inventory reduction. In essence, both systems require a vendor to stock a predetermined selection of items used by the company, which can then be ordered at the buyer's convenience. This is the basic concept of *stockless purchasing*, which is itself an outgrowth of systems contracting. In *systems contracting*, the object is not only to avoid having to keep a central inventory of stock items, but also to relieve the buyer of the chore of ordering the material at all. In actual practice, the buyer surveys the need at each plant location for repetitive MRO items, establishes the approved list, negotiates with the necessary suppliers to set the estimated quantities, firms up the prices, provides the plant with a simplified order form, and decides who at the plant may place orders. The people authorized to order are usually the local purchasing coordinator, the office manager, and often the stockroom managers. In this way the company avoids carrying the inventory, reduces the paperwork at headquarters, and eliminates the hazards of having local purchases made by unskilled personnel.

Dataphone® purchasing: The Dataphone® is used to speed up the ordering process for items held in stock by the vendor. A communication system made available by the Bell Telephone System and IBM, the Dataphone® permits the fast, accurate, and economical transmission of orders between buyer and seller.

The traveling requisition: A device generally used for the ordering and control of repetitive items, the traveling requisition is designed to permit the department ordering the item to activate the requisition, process it, and send it to a source designated by purchasing. It functions in the same way as an order.

The cash purchase order system: This purchase order is so constructed that part of it is a check. When the order is written, the check is made out in the vendor's name and signed by the appropriate company personnel, usually the buyer. Upon shipment of the goods, the vendor fills in the invoice section of the check, determines the costs, figuring all discounts (cash and quantity), arrives at a net sum, fills it in on the check, and deposits the check.

2. Other Tools and Procedures

The quotation: This is a basic part of purchasing and can be a vital record, depending on how the purchasing department operates. If bids are a large part of the business, then it would be wise to standardize the system for eliciting and receiving bids. There is usually a checklist that the buyer can use to review a bid request before it goes out, making sure it is complete in every detail. If the bid system is to be used, it is vital that each supplier bid on exactly the same set of specifications.

Once the bids begin to be received, there must be a method of recording them for the permanent record and for evaluation. The bid evaluation procedure should be planned in advance. If the bid is a simple one for one item, a *quotation card* can be used as a quick and easy permanent record.

The invoice: In any purchasing operation, the buyer will eventually see the bill for the goods that he has purchased. As far as the invoice is concerned, the buyer must make sure that the invoice is correct and take remedial action if it is not.

The purchase-history card: This is used for recording purchase data on both repetitive items (in which case it becomes an interim record) and one-time buys. The purchase-history card is an excellent device for the buyer to use for reviewing the entire purchase history of an item.

The vendor list: This may merely be a list of vendors presently being used or may contain the names of potential suppliers that have been checked out and found suitable for future use.

The correspondence file: This should be set up as simply as possible in order to facilitate filing and make it easy for someone not familiar with the files to find needed information. A system should be established whereby the interoffice and outside correspondence are handled in a manner as simple as the complexities of the business will allow.

The item file: This is the place where all the essential data pertaining to a given item are stored. Everything pertaining to the item, such as the correspondence, quotations, specifications, and quality control reports, goes into this file. Among the many advantages of this system is the fact that copies of the purchase orders and all of the information necessary to prepare future orders are at the buyer's fingertips. The use of computers has increased the trend toward identification of items by number, which, in turn, has greatly reduced the chances for error.

3. Application of Electronic Data Processing

The computer is a fact of purchasing life and is clearly here to stay. Therefore, the buyer must learn to harness the computer. At first glance this may seem an impossible task, since the average buyer is generally not grounded in the technical aspects of computer programming; but difficult though it is, it is not impossible. First, a new buyer should attempt to overcome the current knowledge barrier that exists between purchasing people and the technicians who run the electronic data processing (EDP) equipment. The next step is to learn a little about the general capabilities of the computer. Once these are understood, the buyer can begin to develop some practical applications for the purchasing department.

EDP equipment covers a wide area, ranging from the relatively simple punch-card systems to highly sophisticated systems that make use of complex computers and supporting equipment. Regardless of the degree of sophistication, it is all used for the same purpose: to provide accurate, timely, usable information. EDP equipment—in all its various shapes, sizes, and speeds—gathers, calculates, and reports data.

The *total integrated system*, often referred to as an MIS or total information system, is one approach to applying the EDP concept to the purchasing function. Here, *system* means a routinized method of satisfying the information needs of management, *integrated* refers to the interchange of information between a home-office system and related plant systems, and *total* means that the system can supply all of the data requested by management.

The computer is a high-speed machine geared to handling large-volume items, and as such is extremely useful as a purchasing tool. If information is to be used in several places, the computer should probably be handling it. Each day the list of things the computer can do grows longer. Buyers should be aware of the capabilities of the computer to relieve them of some of the routine work, help them make decisions by providing comparative data, and free them for more important duties. Areas for EDP application are summarized below:

1. Preparation of automatic requests for quotations
2. Updating of order records by entering progress information
3. Routine follow-ups
4. Requisition processing
5. Checking history files and collation of data on vendor performance to help in the vendor selection process
6. Trafficking (recording receipt of goods and providing the buyer with a notice of the incoming goods)
7. Price files maintenance
8. Preliminary selection or recommendation of supplies on the basis of the buyer's criteria (probably a combination of quality, price, and service)
9. Preparation of cost-control reports
10. Calculation of buying requirements on the basis of production and sales forecasts
11. Calculation of economical order quantities
12. Inventory (writing purchase orders, receiving documents, providing inspection reports, requisitions, and accounts payable information, and calculating the inventory situation at the end of each day)
13. Matching invoices and orders and preparing the accounts payable checks

Although the potential for computer use is limited only by the needs and imagination of the buyer, there is an important drawback to EDP: The more work is converted to mechanical routines, the more flexibility is lost. Clearly, the computer does not think for itself. The buyer must carefully feed it the correct information in order for it to produce the expected and desired results. Too often, people expect miracles from the computer and are disappointed when the hoped-for results do not materialize.

VII. MARKETING

Marketing includes all operations of a business system that are involved in determining and influencing the demand in the market place and activating the supply of goods and services; in other words, marketing equals sales.

With corporate success so often synonymous with marketing success, it is understandable that managements are particularly concerned with the quality of decision making in this key functional area. Fortunately, the marketing decision maker now has access to management science techniques for use in the areas of marketing research, pricing, promotion and advertising, and sales.

A. Marketing Research

Marketing research includes all research involving marketing institutions, marketing function, marketing cost, or marketing policy. It applies equally well to information obtained from the field via consumer questionnaires and to analyses of quantitative and qualitative information available in published sources or in the files of the business enterprise, as long as the purpose of the investigation is that of adding to the fund of current marketing knowledge.

In recent years the number of marketing researchers has increased greatly, and the scope of their interests and activities has become wider and more diversified. Whereas early marketing research tended to be confined to market characteristics and to analyses of selling effort, today's marketing research may deal with buying policies, merchandise handling, location of facilities, transportation costs and methods, risk management, credit and finance, product development and testing, and general marketing administration.

The nature of contemporary marketing research activity has also been influenced by its increasingly important role in management decisions. There are two distinct reasons for this development.

1. Managers have often insisted that research specialists develop information as a substitute for intuition. A manager who has consistently and obviously made mistakes in intuitive decisions in a certain area may request help from a researcher in this area. Also, a manager

who is inquisitive and dissatisfied with the general available knowledge on a certain topic may seek the assistance of a research specialist. There have been occasions when managers have insisted that researchers develop measurements that could not yet be made because of limitations in research techniques. Although the progress of research has been occasionally impeded and its reputation soiled by such demands, the overall effect has been one of greater progress when managers themselves have called for new kinds of information and encouraged research in previously unexplored areas.

2. Research specialists have frequently, without managerial prompting, pioneered the development of new kinds of information for management. There are broad areas in which managers were formerly content to depend solely upon intuition in making marketing decisions but have come to depend upon research because the researcher has developed new knowledge that supplies the basis for better-informed decisions. The ability of researchers to sell aggressively not only the concept of research as a substitute for intuition, but also specific research procedures and projects, has been one of the key factors in the rapid growth of the research function.

The course of research and its progress has also been affected by the degree and kind of interaction that has occurred among researchers themselves. The state of a scientific discipline inevitably depends upon communication among those concerned with that discipline. In brief, the progress of marketing research has been influenced by the acceptance of the consumer-oriented marketing concept, by managers requesting information, by researchers suggesting areas of application, and by growing professionalism among research specialists.

1. Consumer Behavior Research

One of the chief concerns of marketing research is the question of how the consumer will behave. If the answer is known, an exact marketing policy can be developed. Of course the degree of accuracy for any prediction regarding consumer behavior is limited.

Consumer behavior research is concerned with individuals in their role as consumers and emphasizes the relation of specific consumer traits and characteristics to the patterns and rates with which a certain product is bought. The following types of study are included in the scope of consumer behavior research.

Consumption rates (of brands and products) *vs social status:* For example, a model describing automobile purchase behavior says that the critical factor in the consumer's make and brand selection is the consumer's self-perceived social status. Customers with a high social status buy completely different makes of automobiles than customers with low social status.

Motivations for purchasing particular products or brands within a product category: There is a general model of human behavior which puts forth the idea that specific and recognizable forces of motivation vary from individual to individual and from situation to situation. Therefore, it becomes necessary to identify the forces at work in a given product category (say, detergent) that might encourage the purchase of a particular product type (powder, liquid, tablet) or brand.

Purchase behavior vs consumer's life cycle position: A model of purchase behavior may state that purchases are critically related to the consumer's age, marital status, and number and age of children, if any. This model hypothesizes that the consumer's style of living and dominant interests are a function of the stages in a distinct life cycle and that his consumption patterns depend upon his or her position in this cycle. Consumer behavior research might be undertaken to confirm this model both generally and in relation to certain product groups. By feeding the model continuous data on consumption by stage in life cycle, changes in general consumption patterns could be plotted.

Consumer spending vs future income expectations: A model of consumer behavior might suggest, for example, that long-term commitments to durable consumer products (such as a new home or car) are highly related to consumer expectation of a permanent increase in income. Consumer behavior research could test such a model and also feed it with projections of total consumer income expectations and measurements of general attitudes toward the soundness of the overall economy.

2. Marketing Research Techniques

a. Data Collection

There is a widespread notion that marketing research is primarily concerned with the collection of information through the conduct of surveys of consumers or business firms. However, by far the greatest part of the quantitative data and qualitative information used in marketing research activities comes from internal, governmental, and private sources. There are three classic methods of marketing data collection: interviews (in person, by telephone, or by mail), observation, and processing of secondary (accounting) records and materials.

Internal information: The types of marketing information generated within the business firm during its day-to-day operations are many, falling basically into three main categories: (1) data about customer characteristics or behavior, (2) information about the distribution or allocation of marketing effort, and (3) facts that would help in evaluating marketing performance. For example, in a department store, detailed information about certain attributes of credit customers is on file in the credit applications department. Such information can be used to construct frequency distributions of customers according to the level of income, age, occupation, and residence.

Opportunities to learn more about the way in which a firm distributes or allocates its marketing effort are also numerous. For example, in a manufacturing firm with a sales organization, data from salespersons' call report sheets can be tabulated periodically to find how many sales calls are made upon various classes or types of customers.

In order to evaluate performance, internal sources of marketing information are often supplemented with data available from external sources. For example, a furniture manufacturer can compare the company's own sales of each type of furniture with comparable industry statistics for total production or sales. In this manner it may be possible to find specific segments of the business in which the firm is gaining or losing out, in terms of relative share of market position, and to pinpoint specific trouble spots that may call for corrective action or improved marketing approaches.

Government sources: In some way or another, government agencies are involved in the performance of all of the marketing functions, so that large amounts of marketing information can be obtained from them. Census information is the chief government source of current marketing facts. Through periodic censuses conducted by the federal government, detailed information is collected from some of the most important sectors of the national economy. Census data are provided for a variety of reasons and can be widely used for a large number of business, economic, political, and other social and individual purposes. From the standpoint of marketing, they supply most of the basic knowledge about the institutional structure for marketing and the composition of national and local area markets.

Private sources: One of the most significant sources of marketing data is the trade or professional magazines or business papers. These publications often provide quantitative information about the size of the market and qualitative characteristics of the consumer units in the market, their buying power, and the kinds of method that can be used to reach them.

Interviewing method: Personal interviewing is the classic method of collecting data for marketing research. In this procedure, an interviewer makes a personal contact with a second person or respondent who has some designated characteristics. In the personal contact, the interviewer obtains information from the respondent according to some specific questions. The questioning usually follows a formal questionnaire that may vary in length and may involve various aids such as pictures, definitions, and actual products to enhance understanding and communication. Telephone interviewing is basically the same as personal interviewing except that contact is made and data are gathered by telephone rather than face-to-face.

In situations where personal or telephone interviewing is not possible or sufficient, data can be collected by mail. Mail surveys depend upon the respondent to fill out questionnaires according to included instructions. The success of mail surveys depends on the literacy of the respondents and upon the degree to which they can be interested and motivated to complete and return the questionnaires.

Observation: The specific behavior of individuals is observed and recorded in this method of data collection. Usually, the action or condition of interest is defined by the researcher before the observation begins. The individuals under observation are usually unaware that they are being used for a source of data. Thus, the data collected are not mediated or colored by the respondents.

Processing available data: This method of collecting data involves the collection of information from secondary sources, that is, accounting records and other business-related materials available to the researcher. In this method, data generated for other business reasons are used for marketing purposes. It is important for the researcher to determine what kinds of datum are routinely available or obtainable and convertable for marketing use. The development of explicit marketing models has created a demand for such secondary knowledge of a firm's activities either directly or indirectly related to the marketing process.

b. Basic Tools

The management scientist working in the marketing area uses four basic tools plus a number of special models. These basic tools are as follows:

Matrix algebra: This could be used in marketing research to find the total hours of sales effort and dollars of advertising expenditure required to achieve certain geographical sales targets. Matrix algebra has many other applications in marketing and affords the advantage of economy in quantitative expression. (See Chapter 3, Section III.C for an example of the use of matrix algebra in Markov chains.)

Differential calculus: Using differential calculus, the management scientist can determine what combination of inputs will maximize some output. The chief contribution of differential calculus to marketing is to enable a direct determination of optimal action where differential functions are involved. In fact, marginal analysis, which is applied by economists to all kinds of decision situations (such as determination of the best price or the number of salesmen), is really an application of differential calculus.

Probability theory: How should the marketing specialist handle the uncertainty that surrounds legislation, consumer intentions, and competitors' actions? Probability theory (Chapter 3, Section III.A) helps put alternatives into relative perspective. For example, one might list all the possible consequences of a business move and their corresponding probabilities of occurrence (decision tree analysis, Chapter 3, Section III.B). The probabilities can be based on either the frequency outcomes of past events for similar business moves or personal judgment. Probability numbers can also serve as weighting factors for appraising various values or utility outcomes (expected value—see Chapter 3, Section IV.D).

Simulation: The great majority of making problems probably will remain unsolvable by ordinary mathematical means. For example, the correct price to charge depends on such factors as the future sales outlook, the possible reaction of competitors, the time lags between these reactions, and the intended level of advertising support. A complex situation is characterized by feedbacks, distributed lags, uncommon probability distributions, and other factors that make exact mathematical solutions very difficult or impossible. However, a feasible solution may be found by simulation (Chapter 6, Section III.B.8). Computer simulation has been conducted on such marketing problems as media selection, department-store ordering and pricing, site location for retail outlets, and customer facility planning in retail outlets.

c. Models

The above-mentioned tools are fundamental in setting up and solving many of the models that have been developed to aid marketing executives in decision making. Some of these models are designed for normative decision making, and others for the analysis of a process.

Allocation models: The economic aspect of scientific decision making is the allocation of scarce resources to competing ends. In marketing, the scarce resources could be salesmen who are too few to make all the desirable contacts, or limited advertising dollars. A decision must be made on how to allocate or program these limited resources to territories, classes of customer, and product lines. This can be accomplished by using some type of allocation model, of which there are several: those using linear programming, nonlinear programming, dynamic programming, and simulation (Chapter 6, Section II.D). These programming models hold great promise for aiding in the solution of such marketing problems as media selection, allocation of sales force, determination of best product line, site location, and selection of channels of distribution.

Competitive models: Profit outcomes are not only a function of the decision made by a firm, but also of this decision made in conjunction with the decisions made by competitors. A marketing decision must be based on an estimate of what competitors are likely to do, even though their decisions may not be known in advance. *Game theory* (Chapter 6, Section III.B.7) is a systematic investigation of rational decision making when the uncertainty of the moves of competitors is involved. Although game models do not seem to have much predictive power, they do suggest a useful analytical approach to such competitive problems as pricing, sales force allocation, and advertising outlays. Another competitive concern, market share, is handled by *brand-switching models:* Marketing executives must watch their market share just as much as their profit. Customers should never be taken for granted. The attitude of marketing executives toward brand switching is, of course, that the switching-out rate must be decreasing and the switching-in rate must be increasing. In order to alter the existing brand-switching rates, the factors affecting brand choice should be analyzed. A *brand-switching matrix* can be constructed, using probability theory, to provide information about each brand's repeat-purchase, switching-in, and switching-out rates.

Queuing models: Queuing theory (Chapter 6, Section III.B.5) is applicable to many marketing situations. Often customers wait for service, and companies wait for both customers and deliveries. Waiting is of interest because although it imposes a cost, so does the effort to reduce waiting time. The decision problem is one of balancing the cost of lost sales against the cost of additional facilities. Queuing theory can aid marketing executives when facing such decisions.

B. Pricing

Price theory was mainly developed to explain and analyze the operation of a market economy. The price mechanism provides an excellent yardstick to measure the efficiency of a competitive free-enterprise system.

Prices are a basic and critical factor in determining profits and ultimately the success or failure of a business. The ultimate goal of a pricing policy is to formulate a program so that over the range of economic conditions the owners will receive a satisfactory return on their invested capital. An effective pricing policy is the result of coordination by management on costs, volume, prices, profit, competition, and market demand.

Theoretical price models assume that other marketing variables are held at some constant level while the effect of price on sales is being examined. In reality, all marketing variables interact to such an extent that optimization is difficult. It is also difficult to predict a competitor's reaction to a price change. For example, the price obtained to generate maximum profits may attract new companies into the market, thereby reducing the long-term well-being of the firm. In the situation of risk and uncertainty, the manager should try to see how sensitive the theoretically calculated price is to revisions in the estimated data.

1. Profit Maximization Model

This model assumes a profit maximizing firm that has knowledge of its revenue and cost functions. The revenue function describes the level of revenue R based on the output sold q by the firm. The cost function describes the expected level of cost C for the quantity sold by the firm.

If we have continuous cost and revenue functions, the profit function is

$$Z = R(q) - C(q) \tag{7-9}$$

The maximum output will be at a stationary point q, where

$$\frac{dZ}{dq} = \frac{dR(q)}{dq} - \frac{dC(q)}{dq} = 0 \tag{7-10}$$

At this point, the marginal revenue $dR(q)/dq$ equals the marginal cost $dC(q)/dq$. The solution of Eq. (7-10) will locate the profit maximizing quantity q^*. The optimum price P^* is then found with $P^* = f(q^*)$. In order to ensure that a maximum, and not a minimum, has been located, a second-order test of optimality is necessary. The condition is

$$\frac{d^2Z}{dq^2} = \frac{d^2R(q)}{dq^2} - \frac{d^2C(q)}{dq^2} \le 0 \tag{7-11}$$

Several restrictive assumptions are involved that severely limit the model's applicability to actual pricing problems. It is assumed that the demand and cost functions can be estimated with sufficient accuracy and that the competitors' reactions are negligible.

2. Markup Pricing

In this method the price is determined by adding some fixed percentage to the unit cost. There are several rules of thumb used in practice in connection with markup pricing. Two of the most commonly followed are as follows:

Markups should vary inversely with unit cost

Markups should vary inversely with turnover

Since this method is founded only on general experience, it is doubtful that it would lead to prices generating maximum profit either in the short or long run. One of the major flaws of markup pricing is that it ignores current demand elasticity. As demand elasticity changes, as it is likely to do seasonally, the markup should also change.

3. Target Pricing

A common cost-oriented approach used by manufacturers is known as *target pricing*, in which a firm tries to determine the price that would give it a specific target rate of return on its total costs at an estimated standard volume of production.

Management's first step is to estimate its total costs at various levels of output. The next task is to estimate the percentage of capacity at which the firm is likely to operate in the coming period. After a target rate of return is decided upon, the price needed to generate the required revenue may be calculated.

Target pricing, however, has one major flaw. The company uses an estimate of sales volume to determine the price, whereas price is a factor that influences sales volume.

4. Contribution Pricing

Contribution pricing is similar in nature to target pricing. Although this method ignores demand elasticity, it can provide guidelines to management upon which to base a more realistic price policy.

There are two basic elements involved in contribution pricing.

Variable costs: Those costs that are directly applicable to a particular product

The profit/volume ratio P/V: The percentage of the sales dollar variable to cover fixed costs (overhead) and profit after deducting variable costs

When the *P/V* is known, it is possible to determine the effect on profits that additional sales will produce. Knowing the *P/V* also allows management to set a price that will just cover variable costs but will contribute nothing to cover fixed expenses.

The important aspect of this method is that it analyzes pricing in relation to fixed and variable costs and shows what percentage of sales revenue goes to cover each category of cost. This information can be important to management when consideration is being given to price changes.

C. Promotion and Advertising

In general, *promotion* is any activity designed to stimulate sales and ultimately increase profit. Such activities include providing information to customers, trying to modify their desires, and supplying reasons to prefer the particular company's brand.

When a firm considers advertising, it faces two major decisions: how much total effort to invest in advertising, and what advertising strategy to use.

In dealing with the first problem, the total advertising budget should be established at a level where the marginal profit from the marginal advertising dollar equals the marginal profit from using the dollar in the best nonadvertising alternative.

After establishing the budget size, the firm has to choose media and decide how the budget should be allocated. Therefore, the firm has to decide how often ads should be used and, if relevant, during what time of the day or night. This is called *media scheduling*.

There are several techniques that can assist the manager to make a sound decision when deciding the size of advertising budget or the allocation of budget and advertising strategy.

The percentage of sales method: Many companies prefer to set their advertising expenditures at a specific percentage of sales. A number of advantages are claimed for this method: Expenditures are likely to vary with revenue, and competitive stability is encouraged. However, this method does not consider the relationship of advertising costs and advertising effects. Furthermore, it deals with the availability of funds rather than the opportunities.

Objective and task method: This method finds the total advertising budget by first defining specific objectives for each product and territory. Costs are estimated for each goal and added to determine the total budget. An advantage of this method is that effort is concentrated on alternatives thought to be most rewarding. However, the method fails to question whether a goal is worth pursuing in terms of its cost. The required advertising expense may be too far out of line with the likely contribution of this objective to profits.

Programming models: Mathematical programming could be used to provide an alternative framework for finding the optimal marketing mix. The amount of sales will depend upon the mix of advertising objectives, for which only limited resources are available. The main objective is to maximize total profit by establishing optimal sales volumes and marketing mixes. First the objective function, which is a statement of marginal contributions of each product to profit, must be determined. Then the marketing requirements and constraints for different sales levels must be expressed mathematically. Once the constraints are stated, depending upon the type of model, appropriate techniques could be employed to solve for the optimal solution—that which lies within constraint boundaries and produces the maximum profit.

The most difficult task in the introduction of the systems approach to marketing and advertising management is to get top executives involved in the concept. This difficulty arises because advertising executives often do not understand the concept, confusing the systems approach with the system techniques. All human-fabricated systems, of which advertising is one, have many similar basic properties that allow the basic concept of general systems theory and the basic techniques of operations research and decision theory to establish a common bond among these systems. Yet sufficient differences remain among each of these so that every system requires its own ideal structure and objectives for which specially tailored techniques must be adopted.

Advertising is unique in that it must bridge marketing and communications to develop a synthesis of the two areas. Advertising also interacts to a very high degree with socioeconomic and cultural systems. Advertising executives must be interested in modeling all the useful communications theories to develop part of their sysems requirements; they especially need models for cost minimization in media selection and for special techniques to further their knowledge of audience pattern, media characteristics, and advertising response functions. On the management side, the use of more sophisticated tools of game and statistical theories or the more realistic use of mathematical programming and simulation is a very effective approach, as is the study of adaptive behavior for control of the advertising system. The management scientist who can meet these demands will be able to influence a field that accounts for nearly 2% of the Gross National Product.

D. Sales

The sales process generates its own data, since many details of individual sales are usually recorded. The data describing sales can be accumulated on a number of bases:

From operating purchases to direct delivery

As a record for billing purposes

As an input into inventory control systems

As a weekly total to indicate personnel and raw material requirements

Most sales records can be analyzed as a basis for understanding and controlling the sales operation itself.

Sales research involves two basic processes:

1. The simple processing of sales statistics concerned with how sales are distributed by state or other geographic area, by type of outlet, by distribution type, or by any other applicable descriptive function
2. The analysis of sales with respect to sales data to gain insight into the future development of the firm

The management scientist should consider the application of models that are responsive to the sales statistics. There are several techniques, such as dynamic programming, break-even analysis, regression analysis, statistical tests, and forecasting models, that could be applied to aid management in decision making in this area. (See Chapter 6, Sections II.B.1.d, II.D.1.b, and III.B.1 for discussion of some of these techniques.)

VIII. MANAGEMENT SCIENCE APPLICATIONS IN INTERNATIONAL SYSTEMS

With the increasing popularity of the international corporation, management science is also developing a major role of growth. It appears that an effective international management information system not only can improve management control over the daily business but also can improve top management's ability to determine future plans and strategies. As world markets become more competitive, market sensitivity becomes essential, and there is a real need for an information system that can be adapted to different market situations and management needs. To fill this need one simply cannot transplant the management information system developed for the domestic enterprise to the international operations. This approach will lead to the obvious problems such as accounting differences, as well as to extensive problems in marketing and finance. Any management information system that attempts effective planning of world markets must have the capacity to provide access to reliable sales performance and other market data, as well as have the ability to analyze the data. In designing and maintaining such a system, the multinational firm finds that the time lag and costs are greater than in the case of the domestic firm because of problems in collecting and interpreting the information. Market share data are available only in the most industrialized countries, for instance, and even such basic economic data as national income and price statistics are not available in many countries for a year or longer after the period in question.

Most importantly, perhaps, in the international sphere, the difficulties in achieving an effective system are compounded by differing environments which are not readily understood by foreign management systems.

It is essential for any management system to portray these environmental differences adequately if it is to have any chance for success.

Another major area of concern to any realistic management information system is that of international finance. A foreign subsidiary's performance cannot be assessed in light of U.S. background information. There is instead a need for regular reporting of economic and political data such as rate of inflation, rate of growth in GNP, interest rates, level of foreign reserves, black market exchange rate, and legal changes. Such information is required not only to assess performance but to forecast such things as cash flows and devaluations. Thus any management information system that will truly meet the needs of the international corporation must be extremely flexible and provide for much data, especia'ly environmental data. The need for such a system grows greater monthly and the development of such a system is a major objective of all would-be international corporations.

REVIEW QUESTIONS

1. Give three examples of how sensitivity analysis can be used in accounting.

2. Give two examples of how PERT can be used in finance.

3. Discuss how management science can be applied to international lending and foreign exchange.

4. What is implied by a break-even point? Explain how it can be useful in finding the profit or loss of a corporation.

5. How can management science be used to solve personnel problems? Give three examples.

6. Discuss how behavioral modeling can be used in employee selection.

7. What are the assumptions made in time series analysis?

8. Diagram a production system and explain what major factors should be considered.

9. Discuss how control charts can be used in measuring the quality of products.

10. Explain the general implications of type 1 and type 2 errors.

11. What is the main objective of process analysis? Explain how this method is used in a production system.

12. Discuss the objectives of motion economy.

13. Define order point, lead time, maximum demand, and safety stock.

14. Discuss the rationale for the derivation of optimum order quantity.

15. What difficulties may be experienced in applying economic lot size models?

16. List operational tools and procedures used in purchasing.

17. Briefly discuss the application of electronic data processing in purchasing.

18. What is the main objective of marketing research?

19. List and explain three models that are used in marketing research.

20. Contrast markup pricing and target pricing.

REFERENCES

1. Summers, E. L., Audit Staff Assignment Problem: A Linear Programming Analysis, *Accounting Review* 47, 443–453 (July 1972).
2. Myers, J. H., and Forgy, E. W., The Development of Numerical Credit Evaluation Systems, *Journal of the American Statistical Association* 58 (303), 799–806 (September 1963).
3. Smith, P. F., Measuring Risk on Installment Credit, *Management Science* 11 (2), 327–340 (November 1964).
4. Mehta, D., The Formulation of Credit Policy Models, *Management Science* 15 (2), B30–50 (October 1968).
5. Liebman, L. H., A Markov Decision Model for Selecting Optimal Credit Control Policies, *Management Science* 18 (10), B519–525 (June 1972).
6. Hester, D. D., An Empirical Examination of a Commercial Bank Loan Offer Function, *Yale Economics Essays* 2 (1), 3–57 (Spring 1962).
7. Markowitz, H. M., *Portfolio Selection: Efficient Diversification of Investments*. New York: Wiley, 1959.
8. NABAC Research Institute, *A Study for Improving Commercial Teller Operations*. Park Ridge, Illinois: NABAC, 1963 Vols. I, II.

RECOMMENDATIONS FOR FURTHER READING

Aaker, David A., Addendum to "Management Science in Marketing—The State of the Art" *Interfaces* (*TIMS*) 4 (4), 38 (August 1974).

Aldeison, W., and Green, P. E., *Planning and Problem Solving in Marketing*, Homewood, Illinois: Irwin, 1964.

An Aware Consumer Rates the Industry, *Merchandising* (April 1976).

Barbosa, Lineu C., and Friedman, Moshe, Deterministic Inventory Lot Size Models—A General Root Law, *Management Science* 24 (8), 819–826 (April 1978).

Barksdale, H. C., and Weilbacher, W. M., *Marketing Research*. New York: Ronald Press, 1966.

Barlow, C. W., *Purchasing for the Newly Appointed Buyer*. New York: American Management Association, 1970.

Bates, W. T. G., A Systematic Approach to Personnel Selection, *International Labor Review* (July 1972).

Bierman, H. and Smidt, S., *Captial Budgeting Decision*, New York: Macmillan, 1966.

Blankenship, A. B., and Doyle, J. B., *Marketing Research Management*. New York: American Management Association, 1965.

Bruce, James W., Jr., Management Reporting System—A New Marriage Between Management and Financial Data Through Management Science, *Interfaces (TIMS)* 6 (1), 54–63 (November 1975, Part 2).

Buchan, J., and Koenigsberg, E., *Scientific Inventory Managment*. Englewood Cliffs, New Jersey: Prentice-Hall, 1963.

Buffa, L. S., *Modern Production Management: Managing the Operations Function*, 5th ed., New York: Wiley, 1977.

Cantor, J., *Evaluating Purchasing Systems*. New York: American Management Association, 1970.

Clark, W. A., and Sexton, D. E., *Marketing and Management Science: A Synergism*. Homewood, Illinois: Irwin, 1970.

Cohen, Morris A., and Pekelman, Dov, LIFO Inventory Systems, *Management Science* 24 (11), 1150–1162 (July 1978).

Corcoran, A. Wayne, The Use of Exponentially Smoothed Transition Matrices to Improve Forecasting of Cash Flows from Accounts Receivable, *Management Science* 24 (7), 732–739 (March 1978).

Crane, D. B., A Stochastic Programming Model for Commercial Bank Bond Portfolio Management, *Journal of Financial and Quantitative Analysis* 6 (3), 955–976 (June 1971).

Daiute, R. J., *Scientific Management and Human Relations*. New York: Holt, Rinehart and Winston, 1964.

Darden, R. W., and Lamone, R. P., *Marketing Management and the Decision Sciences: Theory and Application*, Boston, Massachusetts: Allyn and Bacon, 1976.

Dawly, D. A., and Joseph, D. M., Using PERT in Accounting Reports, *Management Science* (July–August 1970).

Debanne, Joseph G., and Lavier, Jean Noël, Management Science in the Public Sector—The Estevan Case, *Interfaces (TIMS)* 9 (2), 66–77 (February 1979, Part 2).

Demiski, J. S., Accounting System Structure on Linear Programming Model, *Accounting Review* (October, 1967).

De Nisi, Angelo S., and Mitchell, Jimmy L., An Analysis of Peer Ratings as Predictors and Criterion Measures and a Proposed New Application, *Academy of Management Review* 3 (2), 369–374 (April 1978).

England, W. B., *The Purchasing System*, Homewood, Illinois: Irwin, 1967.

Fine, I. V., Westing, J. H., and Zenz, G. J., *Purchasing Management: Materials in Motion*. New York: Wiley, 1969.

Gravereau, V. P., and Konopa, L. J., *Purchasing Management: Selected Readings*. Columbus, Ohio: Grid, 1973.

Greenlaw, Paul S., Management Science and Personnel Management, *Personnel Journal* 52 (11), 946–953 (November 1973).

Hanssman, F., *Operations Research in Production and Inventory Control*. New York: Wiley, 1962.

Harris, R. D., and Maggard, M. J., *Computer Models in Operations Management: A Computer Augmented System.* New York: Harper and Row, 1972.

Heinrits, S. F., *Purchasing.* New York: Prentice-Hall, 1947.

Hertz, D. B., Risk Analysis in Capital Investments, *Harvard Business Review* (January–February 1964).

Holstein, William K., and Berry, William L., The Labor Assignment Decision— An Application of Work Flow Structure Information, *Management Science* 18 (7), 390 (March 1972).

Horngren, C. T., *Cost Accounting: A Managerial Emphasis.* Englewood Cliffs, New Jersey: Prentice-Hall, 1972.

Jones, G. A., and Wilson, J. G., Optimal Scheduling of Jobs on a Transmission Network, *Management Science* 25 (1), 98–104 (January 1979).

Klastorin, T. D., On the Maximal Covering Location Problem and the Generalized Assignment Problem, *Management Science* 25 (1), 107–112 (January 1979).

Kollios, A. E., and Stempel, J., *Purchasing and EDP.* New York: American Management Association, 1966.

Komar, R. I., Bank Liquidity Management Viewed as an Inventory Problem. Paper addressed to the Operations Research Society of America, 41st National Meeting, New Orleans, Louisiana, April, 1972.

Kotler, P., Marketing Mix Decisions for New Products, *Journal of Marketing Research* (February 1964).

Leterman, E. G., *Sale and Sales Management.* New York: Alexander Hamilton Institute, 1970.

Lynwood, J. A., and Montgomery, D. C., *Operations Research in Production Planning, Scheduling and Inventory Control.* New York: Wiley, 1960.

Mairs, Thomas G., Wakefield, Glenn W., Johnson, Ellis L., et al., On a Production Allocation and Distribution Problem, *Management Science* 24 (15), 1622–1630 (November 1978).

Manes, R. P., A New Dimension to Breakeven Analysis, *Journal of Accounting Research* 4 (Spring 1966).

Mao, J. C. T., *Quantitative Analysis of Financial Decisions.* New York: Macmillan, 1969.

Migginson, L. C., Personnel Behavior Approach to Administration, in *Personnel Management.* Homewood, Illinois: Irwin, 1967.

Miner, J. B., *Personnel and Industrial Relations: A Managerial Approach.* New York: Macmillan, 1970.

Montgomery, D. B., and Urban, G. L., *Management Science in Marketing.* Englewood Cliffs, New Jersey: Prentice-Hall, 1969.

Moyer, M. S., Management Science in Retailing, *Journal of Marketing* 36 (1), 3–9 (January 1972).

Naddor, E., *Inventory Systems.* New York: Wiley, 1966.

Nahmias, Steven, and Wang, Shan Shan, A Heuristic Lot Size Reorder Point Model for Decaying Inventories, *Management Science* 25 (1), 90–97 (January 1979).

Odiorne, G. S., Personnel Administration by Objectives, in *Personnel Management.* Homewood, Illinois: Irwin, 1971.

Pegram, R. M., *Purchasing Practices in the Smaller Company.* New York: The Conference Board, 1972.

Rago, L. J., *Production Analysis and Control*. Scranton, Pennsylvania: International Textbook Co., 1963.

Rappaport, A., Sensitivity Analysis in Decision Making, *Accounting Review* (July 1967).

Riggs, J. L., *Economic Decision Models for Engineers and Managers*, New York: McGraw-Hill, 1968.

Schmidt, J. W., and Taylor, R. E., *Simulation and Analysis of Industrial Systems*. Homewood, Illinois: Irwin, 1970.

Schwartz, G., *Science in Marketing*. New York: Wiley, 1965.

Sharon, Ed M., Decentralization of the Capital Budgeting Authority, *Management Science* 25 (1), 31–42 (January 1979).

Straus, G., and Sayles, L. R., *Personnel: The Human Problem of Management*, 3rd Ed. Englewood Cliffs, New Jersey: Prentice-Hall, 1972.

Trueman, R. E., *An Introduction to Quantitative Methods for Decision Making*. New York: Holt, Rinehart and Winston, 1974.

VanHorne, J. C., *Financial Managment and Policy*. Englewood Cliffs, New Jersey: Prentice-Hall, 1971.

Webster, Frederick E., Jr., Management Science in Industrial Marketing, *Journal of Marketing* 42 (1), 21–27 (January 1978).

Wichern, Dean W., and Jones, Richard H., Assessing the Impact of Market Disturbances Using Intervention Analysis, *Management Science* 24 (3), 329–337 (November 1977).

Wright, G., Breakeven Analysis and Return on Capital, *Accountant* (Great Britain) 163 (9 July 1970).

Wright, J. S., and Warner, D. S., *Advertising*. New York: McGraw-Hill, 1962.

Chapter 8
Minor Management Science Applications

INTRODUCTION

The era of gross and monumental advancement in scientific and technological knowledge has created a new world of responsibility for the manager, whose job has not become easier with time, but more difficult. However, the advent of high-speed electronic computers and the subsequent birth of management science has added greatly to the manager's library of decision-making information.

This chapter provides a concise guide for the management scientist's problem-solving efforts in the research and development (R&D) field. A short preliminary discussion is included to familiarize the reader with R&D, its nature and objective, and the requirement for effective management of the R&D effort by concerned agencies. The bulk of the emphasis is on the application of management science tools and techniques to the solutions of research and development problems. These tools/techniques are discussed under general headings in rather broad categorizations.

I. RESEARCH AND DEVELOPMENT

Research and development, today, is the main factor for growth and progress in all industries; but what exactly does it mean? The National Science Foundation [1] has offered a definition for use in business and industry. The basic definition given by Villers [2, p. 4], follows:

> Research and development includes basic and applied research in the sciences (including medicine), in engineering, and in design and development of prototypes and processes. It does not include quality control, routine product testing, market research, sales promotion, sales

service, research in the social sciences or psychology, or other non-technological activities or technical services.

Related aspects include the following:

Basic research is concerned with exploring the unknown. The primary mission is that of acquiring knowledge for its own sake. Its results are usually unpredictable. The reservoir of knowledge so acquired then provides the groundwork for further applications that are used to produce economic growth and material progress. There are rarely schedules to follow for specific accomplishments since no goals are established. Each new step is planned and chosen based on the results of the previous step. Basic research is the nucleus of future progress.

Applied research is an application of basic knowledge directed toward a specific objective, such as the development of a new product, process, or material. Applied research also includes the improvement of products and processes already in use. Because of its importance to industry, by far the larger portion of all research is in the field of applied, or directed, research.

Development is the process of transferring research results into usable products or services. This definition specifically excludes the routine activities that provide services and products to users. Development contains most of the current management science tools and techniques discussed in this chapter.

Another commonly used term that will help clarify the concept of R&D is *research and development plant*. This is defined as land and capital goods, fixed equipment, buildings, laboratories, testing sites, and other facilities required for the actual research and development functions. The objectives of research programs and the logistic support required or used are excluded from this definition.

Research and development, by nature, is characterized by both high costs and long-term investments. Usually there is a long time-lag involved, five years or more, between the time an idea is initiated in research departments and full-scale production. This time-lag is often traced, not to an actual length of time between conception and production, but to the time that includes all the failures and restarts. It is estimated that 60%–90% of the new products of basic research are failures. This failure does not necessarily occur on the market but may happen during R&D in any of its phases for a variety of reasons. Management, or rather the lack of it, seems to be the reason offered by most researchers why the failure rate of new ideas is so high.

The objective of R&D in any organization is to evaluate new ideas and gain further knowledge of the area under study via systematic methods. Knowledge gained by research is then applied toward the production of new and/or improved products. Some specific objectives of the research and development effort are the following:

1. Discovery and development of new products and knowledge
2. Improvement of existing products and production systems
3. Study of other related industries and their products
4. Provision of in-house studies and services to functional departments

The overall industrial R&D purpose is to increase and maintain company profit.

As mentioned earlier, the fundamental nature of research and development is one of high initial costs with an unknown return in the future. These high initial costs demand the success of new products and processes. Failure or breakdown in any phase can result in excessive losses to a company, both in time and money. Failure in any phase of the R&D spectrum can generally be traced to ineffective management somewhere in the process. Without an effective management program, the R&D program can become like a bottomless pit, consuming huge quantities of a company's resources without producing any viable contribution.

The selection of projects is highly dependent upon the degree of risk and resources required. This selection is a management function that can be aided by management science. Budgeting the resource allocation is another critical area, and this too is a managerial task for which management science tools and techniques can be used effectively.

The establishment of clear objectives and provision of adequate policies are expected of the R&D manager and are essential to an effective R&D effort. Management science does not establish the objectives or substitute for advance planning; however, it does provide alternatives and most probable facts that will allow the concerned manager to make these decisions based on better knowledge. In fact, management science will force the manager and staff to research, collect data, and investigate the environment surrounding the R&D effort, which can only benefit the management process.

Management of research and development entails planning and control: proper decisions at the right time about the right things. It is management science tools and techniques that provide the structure and method for planning. Quantitative methods are extensively used since they yield results that are immediately tangible and readily comparable to other alternatives. Whether the methods are known as operations research, management science, cost effectiveness, or systems analysis, the objective is the same: to allow the manager to view the various alternatives and their subsequent effects in order to produce a better decision.

Those areas that require the most extensive use of management science tools and techniques are *project selection, project management and control,* and *project evaluation.* The remainder of this chapter addresses specific tools and techniques as they may be applied to component problems or decision areas within these broad topics.

II. MANAGEMENT SCIENCE IN R&D APPLICATIONS

In the applied stages of the research and development process, the manager is faced with the problem of translating the research results into products or processes. This transformation takes place in the environmental context of limited resources. The problem of resource allocation is coupled with the problem of deciding among various competing methods of obtaining the desired transformation. The result is that the R&D manager is faced with the problem of determining a relative course of action and then attempting to optimize the allocation of resources within the constraints imposed by that course of action.

This section presents some of the current tools and techniques available to R&D managers.

A. Linear Programming

Allocation problems are good candidates for linear programming (Chapter 6, Section II.D.1.a) whenever the variables in the problem are linearly related to one another. The technique of linear programming has been designed to aid decision makers in decisions involving the distribution of limited resources among a set of conflicting demands, providing a relatively easy computational tool.

In resource allocation, the manager must establish an objective based on the relative return or importance of the various competing variables in the problem. This information can be furnished by historical data from similar development projects, estimates by management, or through the use of weighted predictive data. In addition to the objective function, a set of constraints imposed by the decision environment must also be taken into consideration. The results provide the scheme for the resource allocation.

Linear programming may also be used under certain circumstances to aid in policy strategy.

B. Nonlinear Programming

Obviously, not all problems can be solved by linear programming since many R&D problems are characterized by nonlinearities in the objective function, constraints, or both. Such problems arise when one or more of the variables takes on a nonlinear function. Expressed in economic terms, this situation may be described as the analysis of constrained maximization problems in which diminishing or increasing returns to scale are present. A situation such as this might well be found in a development program where the return from allocated resources tends to increase at a higher rate than the direct application of those resources.

Unlike linear programming, there is no general algorithm for handling nonlinear models, since the feasible solution regions are not as well behaved as in the linear situation. Several techniques that can be used by managers in dealing with nonlinear programming problems were presented in Chapter 6 (Section II.D.1.b).

Nonlinear programming, like linear programming, provides the manager with a method for determining an optimal pattern of action within the environment; however, nonlinear programming problems are considerably more difficult to develop and to solve. As a result, a manager may find that the nonlinear model in use cannot be solved by existing techniques. In such a situation, a solution can best be obtained by dividing the problem, if possible, into solvable subproblems. Even so, the application of the nonlinear programming system will provide the R&D manager with an opportunity to examine the various interrelated factors of the problem. This opportunity may well be worth the time spent in developing the programming model.

C. Dynamic Programming

The research and development manager is faced with the problem of translating the result of the research into a definite product or process, and thus must deal with a host of process-concept, process-control, and alternative-evaluation problems in addition to the problems of planning and managing the development work. Many of the problems facing the manager are of a staged nature; that is, decisions must be made in sequential order, with the output resulting from one decision becoming the input to others. Problems of this nature occur in all development work, whether involving rocket systems or pilot oil refineries. The problem is to find the sequence of decisions that will yield the optimal overall return. This problem may take various forms:

What work station layout will yield the best flow?

What reliability allocation will yield the best overall reliability?

What resource allocation will yield the optimal system return?

A technique that could aid greatly in solving problems of this nature is dynamic programming (Chapter 6, Section II.D.2.b).

Consider the problem of setting up the reliability relationship of the components making up a missile system. The objective is to maximize the overall system reliability and the problem is to determine the optimum component configuration from among several available alternatives. Assuming a serial configuration, this problem becomes one of staged decision making. Each component becomes a link in the chain, as it were, and the configuration of each component is optimized with respect to the preceding components. This problem is typical in that it is almost impossible to solve by enumeration of all alternatives since even a small system can have a sizeable number of alternatives. Dynamic programming provides a means of attacking such problems. Through the concurrent use of digital computers, reasonably quick solutions may be obtained.

Dynamic programming provides the manager with several useful functions. From the problem-solution standpoint, dynamic programming may provide a solution to complex sequencing problems, providing an optimizing scheme of application. From the program-management standpoint, the technique requires the manager to examine the various available alternatives and assess the return associated with each. This process requires a systematic layout of the system, thereby providing an excellent problem-spotting technique. Although dynamic programming may often be difficult to apply, its logic is certainly an aid to any R&D manager.

D. PERT/CPM

Perhaps the most common class of management problem is that which can be expressed as follows: How does a manager schedule activities and resources over time to obtain the best program performance? This problem affects all areas of a project and must be considered of prime importance to the overall program success. The most well-known systems engineering techniques, PERT and CPM (Chapter 6, Sections II.B.2 and III.B.4), have both been used quite extensively in the scheduling and control of R&D projects. In fact, these techniques are perhaps best applied to R&D since such projects are of a one-time nature and usually entail a large number of interrelated activities.

The PERT/CPM network provides a visual representation of those activities that make up the project, allowing the manager to apply resources at the best points and to slip activities that have slack. It also provides a method for making trade-offs of time and cost in such a manner that the maximum advantage is obtained. This dual technique is perhaps one of the most valuable and easiest applied of the management science techniques.

E. Input–Output Analysis

Due to the growing trend to distinguish between the research and the development functions, the funding for these areas also tends to be separated. This separate funding for each phase of a program provides increased flexibility, but it also poses an increased problem for the development management. Therefore, the R&D manager must be vitally concerned with the cost of the development program.

The development manager is also faced with the problem of how much money to spend in order to obtain desired output. Elements such as quality, safety, reliability, and durability are extremely difficult to measure, and yet in highly complex development programs they become extremely important. The problem is how to be assured that what is needed is being obtained at the lowest possible cost.

Another facet of the development management problem is how to decide the overall contribution of individual element outputs, both tangible and intangible, to the total value of the system. This problem of tangible or intangible factors is further complicated by the fact that much R&D is performed by the government or the military, which do not express the same tangible desired results as does private industry.

The technique most applicable to these considerations is known as *input–output analysis*. Input–output theory assumes that the constant returns are to scale and that the input ingredients have to be combined in fixed and stable proportions, that is, that there is no substitution possible among them. The relationship could be stated as

$$x_{ij} = a_{ij} x_j, \qquad i, j = 1, 2, \ldots, n \qquad (8\text{-}1)$$

where x_{ij} is the amount of input i in area j, and x_j represents the amount of output in division or industry j, and a_{ij} is a technical factor.

From the above relation, an input–output model of the system can be derived that takes into account various limiting factors, such as money, labor, and material. The model then provides a method for analyzing situations where little concrete knowledge is available.

This technique forces identification of mission objectives in quantitative terms and provides a basis for evaluating trade-off alternatives for varying inputs and outputs. It therefore provides management with an approach to identifying its objectives and evaluating its ability to accomplish these objectives effectively.

F. Forecasting and Prediction

A significant amount of effort in the applied phases of the R&D program is concerned with the forecasting of future occurrences. One area of cost and materials allocation, for example, would involve the prediction of future demand for these items by the various agencies within the organization. Forecasting may also be utilized to determine the timing of specific program elements.

Some of the available techniques for forecasting were introduced in Chapter 6 (Section III.B.1). Although these are far from being all the various available prediction techniques, they should serve to illustrate the intent of the formulation. Under forecasting or prediction one bases the estimation of future occurrences on historical data or on data generated by similar elements or similar environmental conditions. Hence, it can be seen that in many cases forecasting will be impossible in R&D, owing to the imperfect (or even nonexistent) data on a unique project. Forecasting comes into play in the determination of a strategy when competing events are known and data about them are available.

When a forecast is required for more than two or three years into the future, it is seldom possible to assess all of the factors likely to affect the research and development program. For this reason, mathematical trend curves (extrapolation) can often be used in place of more exact techniques. The R&D environment is highly dynamic in nature, and prediction techniques must be used with this in mind. These techniques do, however, provide a basic framework on which management may build.

G. Expected Value

Decision among competing alternatives is one of the most frequent problems faced by any manager. During the several phases of the R&D project, this problem takes on added complications due to the very nature of a new idea: that there is often little defined information with which to work. Areas such as process concept definition, expected returns, and expected cost must all be considered in the selection of an alternative method of development. One of the most frequently used tools is the expected value method.

The *expected value* of an alternative is the weighted average of all the possible outcomes of the alternative (cf. decision tree analysis, Chapter 3, Section III.B). The outcomes of the alternative are weighted according to their probability of occurrence. Therefore, the first step in developing an expected value decision is to determine the outcomes of the various courses of action. There seldom occurs the alternative that has only one consequence since alternatives are always subject to the two base states of nature—success or failure. However, it is more likely that an alternative will result in a partial success or a partial failure. Although the consequences of the outcomes of an alternative are most often expressed in monetary terms, they may be expressed in any common comparison units.

The second phase of the expected value analysis concerns the probabilities associated with each of the alternative's outcomes. These probabilities come primarily from two sources: the data from previous similar programs and the subjective evaluation of the individuals involved in the program.

A *payoff table* is the standard format for organizing and displaying the outcomes of alternative courses of action. For this, it is necessary to determine all alternative courses of action and the probability of occurrence of each state of nature (see Chapter 3, Section I). In general, the standard format for an expected value computation is as shown in the table of expected payoff values, below:

	State 1	State 2	$\cdots$	State J	$\cdots$	State m
Probabilities	$P(1)$	$P(2)$	$\cdots$	$P(J)$	$\cdots$	$P(m)$
Alternative 1	O_{11}	O_{12}	$\cdots$	O_{1J}	$\cdots$	O_{1m}
Alternative 2	O_{21}	O_{22}	$\cdots$	O_{2J}	$\cdots$	O_{2m}
$\vdots$	$\vdots$	$\vdots$		$\vdots$		$\vdots$
Alternative i	O_{i1}	O_{i2}	$\cdots$	O_{iJ}	$\cdots$	O_{im}
$\vdots$	$\vdots$	$\vdots$		$\vdots$		$\vdots$
Alternative n	O_{n1}	O_{n2}	$\cdots$	O_{nJ}	$\cdots$	O_{nm}

The expected value for each alternative is calculated as

$$E(\text{Alternative } 1) = [O_{11} \times P(1)] + [O_{12} \times P(2)] + \cdots + [O_{1j} \times P(J)] + \cdots + [O_{1m} \times P(m)]$$

$$E(\text{Alternative } 2) = [O_{21} \times P(1)] + [O_{22} \times P(2)] + \cdots + [O_{2j} \times P(J)] + \cdots + [O_{2m} \times P(m)]$$

$$\vdots$$

$$E(\text{Alternative } i) = [O_{i1} \times P(1)] + [O_{i2} \times P(2)] + \cdots + [O_{ij} \times P(J)] + \cdots + [O_{im} \times P(m)]$$

$$\vdots$$

$$E(\text{Alternative } n) = [O_{n1} \times P(1)] + [O_{n2} \times P(2)] + \cdots + [O_{nj} \times P(J)] + \cdots + [O_{nm} \times P(m)]$$

where O_{ij} represents the outcome of alternative i at state j.

Once the expected values for all alternatives are calculated, the alternative with the highest expected value would be assumed to be the best alternative. Similarly, the alternative with the least expected cost would be assumed to be the best alternative.

Expected value provides the manager with a computationally easy method for comparison among various alternatives. It should be recognized that the values placed on both the consequences of each outcome state and the probabilities associated with these outcome states will be somewhat subjective during the basic and applied research phase of the research and development project. However, the benefits gained through a layout of the possible outcomes would itself justify the use of the technique. The ease of use and of presentation makes this technique an extremely viable management tool.

H. Utility Theory

Research and development management's decisions are deeply affected by the characteristics of the decision makers and the decision process in the organization. This aspect of decision making is extremely important during an R&D project because in the majority of the cases there is not a great deal of previous data upon which to base one's decisions. This aspect of the effects of human evaluation decisions is best handled by utility theory.

The *utility* of an outcome is a function of one's view of its usefulness and value to oneself at a given time and place. Therefore, utility is not a constant value but is continually variable with respect to the individual and the circumstances of the moment. It would therefore seem correct to assume that the determination of the utility of a given outcome for a given individual is extremely difficult, and such is indeed the case. Furthermore, although it is clear that at different points in time one may obtain different measures of utility, it is extremely difficult if not impossible to isolate the causes of these deviations. Let us first consider how one would manipulate utilities once they have been obtained and then consider some of the factors to take into consideration in determining these factors in a project research or development situation.

There are two types of utility: marginal utility and expected utility. *Marginal utility* has to do with the change in utility with respect to some independent variable such as time. *Expected utility* refers to the amount of satisfaction that is derived from knowing the probability of the possible outcomes.

Expected utility takes into account the reaction of the decision maker to the risks involved in the decision. A utility index is usually developed on the basis of the monetary compensation. The decision maker would demand to be relieved of the obligation stated in a proposition. This index is then used to modify the expected values of the alternative courses of action. The expected utility of the various alternatives can then be calculated in the same manner as the expected value. The expected utility expresses the nature of the decision maker in conjunction with the consequences associated with the various alternative outcomes.

There are several factors present in the development environment that may directly affect the utility that various alternatives hold for an individual. These factors are generally either technical or economic.

Technical factors include the state of the specific art and related arts at the time that alternatives are being considered. Some of the areas of development require more time and technology and have payoffs over longer periods. Therefore, the technical situation, not only of the moment but also of the future, would affect the manager's utility values.

Economic factors are also extremely important during the various phases of the research and development program. The line of business will often affect the economic situation. Some areas can count on rapid returns from any breakthrough, whereas others may gain nothing for a considerable period of time. The amount of capital required to develop an idea also constitutes an important factor in the utility function of the decision maker.

Utility is a necessary part of any real-life decision process. In research and development, the utility placed on alternative courses of action may well be of equal importance to the numerical returns expressed by those alternatives. Therefore, utility theory provides an effective means of introducing the human element into the decision process.

I. Industrial Dynamics

This tool/technique of management science has gained considerable application in research and development usage. The broad applicability of this technique is due to the project nature of most R&D activities. The crux of an understanding of the R&D process lies in knowing those factors that influence the life of an R&D project.

Industrial dynamics deals with the time-varying (dynamic) behavior of industrial organizations. It is the study of the information/feedback characteristics of industrial activity to show how organizational structure, amplification (in policies), and time delays (in decision and actions) interact to influence the success of an enterprise. This technique examines the interactions among the flow of information, money, orders, materials, personnel, and capital equipment, and is a quantitative and experimental approach for relating organizational structure and corporate policy to industrial growth and stability.

The application of industrial dynamics to R&D is characterized by a number of interrelated stages:

1. Describe the theory of cause and effect interaction and identify the decision policies of the system under study
2. Develop a mathematical model of the system
3. Anticipate the behavior of the modeled system using simulation techniques to investigate the dependability and accuracy of the R&D model
4. Revise and update the model as necessary until the system requirements are satisfied

These steps indicate a method of approach to an extremely complex problem: the overall systems analysis of a complex research and development project. The strength of industrial dynamics in attacking this problem lies in the development of a model that identifies as well as possible those areas that influence the output of the R&D program. Such a model is by necessity extremely complex. The only hope for successful utilization lies in the use of simulation with the model as a base.

From the basic model of the R&D project one may extend this technique to cover other areas of management interest such as multiprojects and competitive situations. The extension provides for additional options that may lead to improved project performance.

Perhaps industrial dynamics should not be called a technique in itself for it is potentially much more powerful. It might be better classified as an extension and expansion of the techniques of simulation into the practical area of management. Still, whatever it is called, it continues to offer much promise as a tool of both material and organizational control.

J. Value Analysis/Systems Analysis

Within the field of research and development a considerable amount of work has been performed on the specialized use of the technique of value analysis. *Value analysis* is concerned with obtaining an optimal value in a given system.

The term *systems analysis* grew out of the application of value analysis to large weapons systems; from our point of view, the techniques may be considered identical.

In the military or industrial R&D program, the optimal situation does not always hinge on the lowest cost for the item in question; the objective may be to maximize the reliability within a certain cost constraint area. Many factors come into play in determining the value of a system. These factors cover a wide variety of fields and it is seldom that one individual will have all the talents necessary to make a successful analysis. For this reason, both value and systems analysis are concerned with a team approach to the problem.

The concept of the R&D project is an outgrowth of this concept of a multidisciplined team. The members of the analysis team are drawn from many varied elements within the organization so that individual experts may examine every aspect of the system. Elements such as reliability, quality, engineering, and design should all be included to ensure an accurate evaluation of the system.

The value or systems analysis consists of a structure of consideration areas that the team is to examine. These consideration areas can often be formulated into sets of questions regarding the system under analysis, such as what is its function and what else will do the function, in order to determine those areas that can increase the overall value of the system.

Systems analysis is usually aimed at producing an optimal design, as opposed to optimizing an existing design (the case with value analysis). The optimization of the design in the early stages of R&D becomes extremely important as the size of the system increases. Minor changes in the design during the development stage of the project often cost more than the value that can be derived from their institution. The same changes in the early design stages may entail only minimal cost, and the return may be significant. Therefore, the timing of the analysis is of extreme importance to the obtainment of a maximum benefit.

As stated earlier, cost is seldom the prime element in R&D evaluations; more often, elements such as design efficiency or reliability will be considered of utmost importance. Therefore, the elements that should be the most likely candidates for improvement are those that contribute the greatest amount to the system value. For example, the element that con-stitutes 40% of the unreliability of a system would be a much more likely candidate for improvement than the element that constitutes only 1%. In large systems one should always consider the number of each type of element to be used [3, pp. 152–154].

Systems analysis, when applied in the early phases of the R&D project, may achieve two major functions: identify those areas where the value of the system may be improved and those where design errors have been made or may be made.

III. OTHER DEVELOPMENTS

This section presents a few of the latest developments in management science tools and techniques that may become significant in research and development applications in the future. Some of these methods or models are little more than well-documented ideas; none can be considered fully operational. A few of these techniques will, perhaps, become well known and freely used; others may never become operational as newer and better methods are discovered.

If one were to attempt to isolate a single most significant characteristic of the nature of management science, it surely would be that of dynamism. It seems that almost every engineering field is highly interested in man-agement science and is developing new tools and techniques of the trade daily. Some of the methods are field bounded, but others show great promise in general applications.

A. Real-Time Computer Scheduling System

This management science tool appears to hold great promise for the future in research and development applications because of its real-time com-puter use. As conceived, the tool provides the manager/management sci-entist with a direct connection to a computer for scheduling and control of research and development projects. The concept is to equip managers (as required) with on-line remote terminals, such as video or verbal input–output machines, to allow them to recall the current stage of the project(s), program changes as they occur, and view the ramifications of those changes, all immediately.

The program is written in MAD, a variation of FORTRAN IV and is a multi-dimensional time-variant method of problem solving. In research and development, it seems applicable to process designing and prototype pro-gramming as well as to scheduling. The program emphasizes the impor-tance of individual decisions on the project quite vividly since there is essentially no time-lag between the decision and its effect on the project.

Actual use of the technique and program in classroom situations has

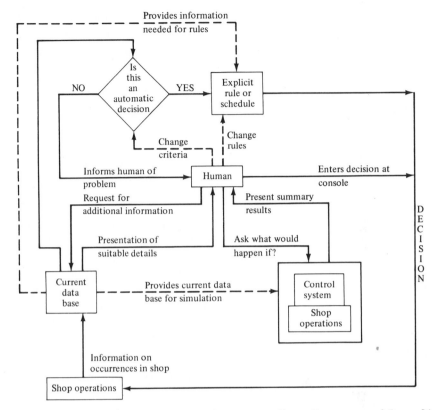

FIGURE 1. A diagram of the scheduling system. *From Ferguson and Jones* [4, *p. 555*].

proved that managers can use the sophisticated computer hardware and the interactive approach to decision making with a minimum of specific training.

A schematic representation of the system is presented in Figure 1. This management science tool, or others of similar capabilities, foretells a new era of human–computer relationships that will certainly be useful in a great number of R&D applications.

B. Dynostat

Dynostat [5] is the name given to a system optimization technique employing both instantaneous and predictive optimum-seeking strategies in parallel. It is intended for use on a high-speed computer and is designed to alleviate somewhat the large core and high computational time required for the more conventional dynamic programming method of optimization.

It is well known that, because of limitations in core storage and available computational time, optimal scheduling by dynamic programming is restricted to consideration of only a few independent variables. The dynostat technique considers (in certain classes of multichannel system) certain variables to be reasonably static; thus optimization of their values need consume little computer time and storage. The remainder of the variables are in the dynamic section and require the expected large core sizes and high computational times. Dynostat handles both types of variable in a single algorithm.

The objective of the technique is to minimize an operation cost function

$$c = \int_0^T \{f(\bar{\sigma}) + g(\bar{x})\}\, dt \qquad (8\text{-}2)$$

where the instantaneous cost functions are resource costs $f(\bar{\sigma})$ and storage costs $g(\bar{x})$. T is the period of time over which the optimum schedule is to operate.

System requirements to be satisfied are the constraints given by the system differential equation and those due to numerous practical limits (i.e., production rates, distributions, etc.). Determination of the optimal schedule requires evaluation of the variables necessary to define the state of the system at any time during the schedule.

The dynostat technique consists of four steps:

1. Quasisteady representation of optimal control trajectories
2. Identification of the distinct sections requiring static or dynamic optimization techniques
3. Application of a repetition of relatively simple, static optimum-seeking techniques within the defined space for each time interval
4. Progressive application of the dynamic programming technique using data from the static optimizer to determine the optimum storage policy from complete trajectories

Using this technique, the computer wastes no time or storage space on uneconomical considerations since they are bypassed by the static optimizer. The digital computer algorithm of dynostat is shown in flow-chart form in Figure 2.

Dynostat, in total, is an amalgam of static hill-climbing and dynamic programming optimization techniques that uses the advantages of both and minimizes the disadvantages. This results in a major reduction in the computation dimensionality problem and exhibits an intrinsic capacity to handle nonlinearities and time-varying coefficients within fairly wide limits.

Dynostat appears to be generally applicable to the same areas of research and development in which linear or dynamic programming are useful, that is, optimization and resource allocation. Perhaps a single dy-

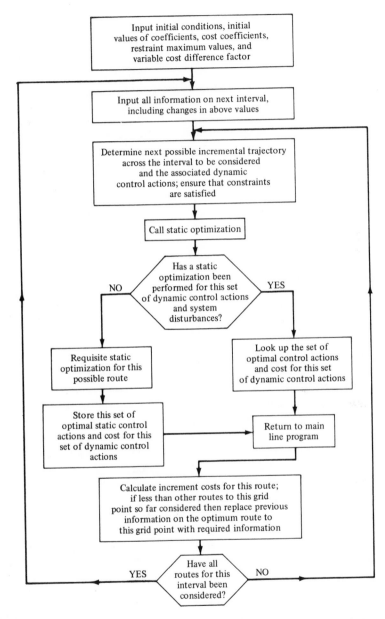

FIGURE 2. Dynostat digital computer algorithm. *From Gibson and Coombes [5, p. 207].*

nostat program available to a firm could replace the three individual tools/
techniques of linear, nonlinear, and dynamic programming.

C. A Model for Research Projects Selection

This management science tool [6] appears to be particularly appropriate
for research and development since it aids in the selection of the projects
that should be continued or started in research. Essentially, the model
is another optimization tool for management science that bridges the gap
between the simple models that treat research as if it were static and the
very complex models that treat it as dynamic. Neither approach is ideal:
The simple models ignore the essential aspects of research, whereas the
more complex methods require so much information that their usefulness
can never be really tested. The model is applicable to optimization and
resource allocation, budgeting, project selection, economic analysis, and
(to some extent) product design and general decision making.

This model is general in scope and has been formulated to apply to
those research projects whose objective is either the invention of new
products or the modification of existing products. The products that are
to be considered are limited to those whose success or failure will be
determined within five years. The technique is based on the development
of a probability model tempered by available commercial information and
the anticipated rate of expenditure. Optimal allocation of resources is then
determined, followed by simulation wherein point estimates of input values are treated as distributions.

The probability model is based on the expected costs of completing a
project (whether already started or merely in the planning stages) in terms
of an optimistic estimate, a most likely estimate, and a pessimistic estimate (as in PERT—see Chapter 6, Section III.B.4). The distribution of
these anticipated costs is treated as logistic with a transformation applied
to allow for skewness. The overall probability of success is calculated by
multiplying together the probabilities of technical, legal, engineering, and
commercial success, since failure in any one of these areas ultimately
means failure for the project.

Commercial information comes from expected sales of the product. The
sales estimates used in the model are those of the first, sixth, and eleventh
years. (The whole analysis is run for an 11-year period. A shorter period
would severely penalize long-range projects, a longer one would induce
excessive forecasting errors.) A correction is applied for those projects
that do not commence on the first of the year or do not emerge from
research and development until later than anticipated. Overhead and selling costs are considered as a constant fraction of sales revenue.

Rate of expenditure is directly related to the probability of success to a certain point. A severe penalty is assessed for late successes, which emphasizes the necessity for adequate expenditure on specific projects rather than across-the-board subsidization of all ideas/projects. Increasing the expenditure of research funds at the beginning of the project results in an increased probability of success in the initial years and so leads to an increased value of the expected profit. Too great an expenditure could result in the wastage inherent in a crash program; therefore, the algorithm defines an efficient annual expenditure for any year. A ratio of the efficient annual expenditure to the greatest cost of completing the project is then formed to measure how well defined the project is. The smaller this ratio, the smaller the amount of the program that will be completed in the first year. Finally, the amount spent in any one year is related to succeeding expenditures, since initial costs will affect the efficient amounts that can be allocated in later years.

This method penalizes expenditures above the efficient level to the greatest extent at the early stages of the project. Clearly, additional funds allocated at this stage will result in parallel exploration with the inherent waste. Later acceleration of the program is apt to be more clearly defined in end use, and thus more justified.

Optimization is accomplished by maximizing the expected return from a set of projects subject to the stated budgetary constraint. This is a formidable optimization problem because of the number of variables and the fact that the linear constraints on the budget in a specific year depends upon and varies with the expenditures of the previous years. The method of optimization is an iterative scheme in which the annual budget is optimized year by year until no further improvement is evident.

The results of the optimization give significantly increased expected returns. This is accomplished by increasing expenditure on the more profitable projects and dropping the less profitable, which seems to be a logical approach.

Simulation then follows, with the input point estimates replaced by distributions, resulting in a distribution of values of the expected return. The distribution of each variable is randomly and independently sampled to establish a particular case, and the payoff is calculated. The average value of the payoff obtained by repeating this process many times is an unbiased estimate of the expected value of the payoff. The probability of success and failure is then presented in a histogram of frequency versus return. This combination of graphical output and optimization results in a powerful tool for describing the outcome of a given pattern of expenditure and for determining what that pattern should be.

This management science tool/technique is an obvious outgrowth of dynamic programming, with technique refinements that greatly increase its applicability in the research field. Certainly, its use will greatly aid management's R&D decision making.

D. The Delphi Procedure

Although the use of the Delphi procedure [7] is not widespread, it is included herein as a state-of-the-art advance that has not yet become fully operational.

The procedure has R&D applications in the decision/problem areas of strategy, planning, and (to some extent) project selection. It is the best current method for long-range forecasting that is not based on specific past events or trends. Certainly, history influences all forecasting of future events to some degree; however, this technique attempts to release the forecast from the shackles of yesterday's trends, cycles, and fixed values. The Delphi procedure may be likened to the old brainstorming technique of forecasting within a definite set of rules and procedures to minimize the effect of group pressure and direct the brainstorming in specific areas to specific problems. It results in forecasts that are explicit, i.e., that state the basis for and the method of the consciously formulated ideas, which are obviously advantageous as compared to implicit forecasts (method of derivation not stated).

The procedure involves the use of a panel of experts in the field(s) of interest. These experts are given a series of questionnaires in which their responses are used to generate the next questionnaire. In particular, the reasons for and against various predictions of the panelists are presented so that the panelists have the opportunity to change their views in response to the arguments of the other panelists. The identity of each panelist, in particular the originator of any given argument, is concealed during the course of the procedure.

The basic problem in the methodology of the Delphi procedure is that of panel selection. This may be particularly difficult if the selectors have little knowledge in the area of the forecast. A suggested technique for selecting the panel is initially to isolate a few experts in the field, query them as to their availability, and request that they suggest possible members. Some names will crop up repeatedly; these should be the panel members.

A secondary problem in the methodology is the wording of questions. Compound questions must be avoided. Follow-up queries should avoid situations where the panelist may agree with the forecast but not the reason for it or vice versa. The use of ambiguous and loosely defined terms must be avoided. Clarity of communication is the key to meaningful results.

The number of questions must be limited. An upper limit of 25 questions seems practical. This restriction allows the panelist to maintain interest in the problem and keeps the goal in sight. Also, the numbering sequence of the questions should not be altered during the cycles. This will allow easy cross-reference to previously submitted questionnaires.

The Delphi procedure permits the systematic use of experts for preparing forecasts in unstructured fields where the relationship of the future to the past is not well defined. It also allows a more in-depth approach, bringing to the fore some of the reasons for a specific expert's prediction. It follows that one of the most important uses of Delphi may be in the development of models and structures by making explicit the implicit models of the experts. Certainly, the procedure will be of great significance in broad areas where no one person can be considered an expert, since it allows combination or consolidation of partial-expert opinions and ideas.

E. Contest

The U.S. Navy's Special Projects Office (of Polaris/Poseidon fame) has developed and implemented a system called Contender Evaluation and Selection Technique (CONTEST) [8] that appears to be extremely useful for choosing among alternatives. CONTEST is a method for organizing, evaluating, and classifying contestants in a competition. The technique facilitates the task of decision making by introducing orderliness of approach, quantification techniques, and mathematical processing to the subjective problem of judging alternatives. The CONTEST method is applicable to the R&D problem/decision areas of decision making, project selection, and product/process evaluation. It clearly has application in any area where there are contending philosophies or competitive circumstances.

CONTEST is designed to be used in the situation where product/process feasibility has been established but the corresponding definitions of product or system requirements remain vague enough that meaningful trade-off analyses cannot be performed; in other words, where conventional optimum-seeking methods cannot be employed to strike an optimal balance in the performance variables.

The technique has two main components:

1. A structure that is easily adjusted to cover the particular subject under consideration
2. A ranking method and processing system for accumulating the scores in a manner consistent with certain assumptions that experience has shown to be "axiomatic"

Phase 1 evaluation consists of posing minimum standards of a "sudden death" variety, that is, each proposal is subjected to a series of gross qualification questions capable of a Yes or No answer. If, in the opinion of an evaluator, a proposal gets a No, the ability of that contender to satisfy the minimum requirements of the program is challenged.

Phase 2 consists of sequential relative assessments of those proposals that have met the gross qualification standards. A block categorization grading system is used with a mathematical procedure for accumulating the effects of the rank ordering, comparing the results from each proposal, and providing an objective technique for successive elimination or the designation of a winner.

The criteria of the technique are such that CONTEST demands a balanced excellence from the contender since the system precludes the effect of overriding superiority in a particular area saturating the evaluation.

The first step in the organization of the competition is to define the significant factors. These factors are then subdivided into their equivalent components (called groups). This defines a hyperspace of N dimensions, where N is the total number of groups assembled under the factors. These N axes cover the entire space of concern and serve as its basis. This technique allows the incorporation of modification as desired and reflects the inherent simplicity of the method.

From the N proposals considered during any iteration, each board member (it is assumed a board of experts has been convened to evaluate the contenders) reviews all N proposals for adequacy on a given question. Each board member then rates that proposal with the greatest central tendency as zero. Based upon this zero assignment, the board classifies the remaining $N - 1$ proposals depending upon whether they fall in the central set (assigned a zero), are clearly superior to the central set (assigned a $+1$), or are clearly inferior to the central set (assigned a -1). The board proceeds in this manner until all of the proposals in each of the groups have been rated. This raw data is then resolved into cumulative $+1$s, 0s, or -1s. The resolution is done for each equal-weight group under consideration. The group iteration consists of proceeding through the groups in all factors, assigning categories to each question, establishing the final category for each group, and then processing the totality of groups, if appropriate.

The general mathematical procedure is as follows. The matrix of questions versus proposals is set up as shown in Table 1 (in this case for factor X, group Y). The Q_{ij} are the classifications of $+1$, 0, or -1 to each proposal package.

The column sum is found by

$$B_j = \sum_{i=1}^{M} Q_{ij} \qquad (8\text{-}3)$$

From these, the mean across the proposals is formed by

$$\bar{B}_{xy} = \frac{1}{N} \sum_{j=1}^{N} B_j \qquad (8\text{-}4)$$

TABLE 1. Factor X, Group Y Matrix

	Proposal		
Question	B_1	$\cdots$	B_j	$\cdots$	B_N
1	Q_{11}		Q_{1j}		Q_{1N}
2	Q_{21}		Q_{2j}		Q_{2N}
$\vdots$	$\vdots$		$\vdots$		$\vdots$
i	Q_{i1}		Q_{ij}		Q_{iN}
$\vdots$	$\vdots$		$\vdots$		$\vdots$
M	Q_{M1}		Q_{Mj}		Q_{MN}
	B_1		B_j		B_N

where subscript xy identifies the particular group of a given factor under consideration.

The variance about the mean is found by

$$D_{xy}^2 = \frac{1}{N-1} \sum_{j=1}^{N} (B_j - \bar{B}_{xy})^2 \tag{8-5}$$

Those proposals that satisfy the requirement $B_j \geq \bar{B}_{xy} - D_{xy}$ have a $+1$ assigned for the group being considered; those satisfying $B_j \leq \bar{B}_{xy} - D_{xy}$ are assigned a -1; and those satisfying $\bar{B}_{xy} - D_{xy} < B_j < \bar{B}_{xy} + D_{xy}$ are assigned a zero. These results are then displayed in a matrix that contains all X factors and Y groups. The total sum of the group is then calculated, the grand mean determined, and the variance established as in the initial procedure. An increment Δ is calculated from

$$\Delta = R/(N-1) \tag{8-6}$$

where R is the range of the total sum. This increment represents the interval between proposals that would occur if the N proposals were uniformly distributed. All proposals that fall below $\bar{B}_{xy} - \Delta$ are eliminated at each interaction.

The designation of a winner is accomplished by the criterion

$$B_{Tj}(\text{max}) - B_{Tj}(\text{max} - 1) > 1.5D \tag{8-7}$$

where B_{Tj} is the total sum and D is the standard deviation of the distribution. Note that this relationship degenerates toward the binary form when $N = 2$; it is thus suggested that a $(+, -, 0)$ evaluation be used, the highest sum then being the judgment criteria.

CONTEST is a relatively simple use of elementary statistics to develop a systematic approach to alternative evaluation where the evaluation criteria are rather vague generalities and not subject to easy quantification.

It seems to be a valuable and appropriate addition to the tools and techniques of the management scientist.

F. TORQUE

TORQUE [9], an acronym for Technology or Research Quantitative Utility Evaluation, is designed to supplement the intuition of managers at all levels by combining expert subjective judgments in a structured fashion to serve as a tool or aid in the decision-making process. It is a framework for quantitatively converting statements about desired future products or capabilities into system descriptions that can then be translated into technological criteria. The methodology is based on the premise that the relative amounts of money spent in various areas of science and technology should reflect the varying degrees of interest in those areas. The problem of allocating funds or achieving balance in the basic and especially the applied research areas becomes a matter of determining how much an additional advance in one field is worth to the firm or agency concerned, as opposed to an advance of equal cost in another field.

A method devised to aid in the preparation of an R&D budget with optimal future payoff should answer the following questions:

What research and development achievements are desired?

When will they be needed?

What is their worth?

What will they cost?

TORQUE supplies answers to these questions by several means:

1. Providing weighted time-phased broad statements of future desired capability or product objectives
2. Providing alternative application options to satisfy these objectives
3. Defining the technological advances required to make these options possible on a timely basis and the importance of these technologies relative to the provision of options
4. Determining the resources required to provide these advances on a timely basis
5. Defining a simulation model that structures the foregoing data to achieve a balanced exploratory development program within any given level of resources available to the firm/agency.

The technique and method of TORQUE is presented in Figure 3 in the form of a flow chart. The responsible agencies and their required contributions are listed therein.

The inputs to the computer are coded time and cost data resulting from the team efforts. The budget allocation is optimized by obtaining the maximum utility from the efforts funded for a given dollar amount avail-

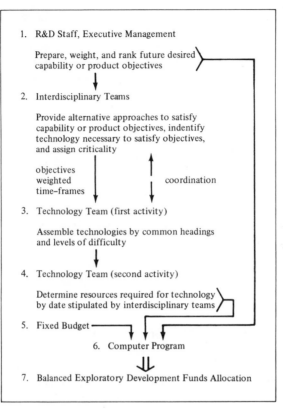

1. R&D Staff, Executive Management

    Prepare, weight, and rank future desired
    capability or product objectives

2. Interdisciplinary Teams

    Provide alternative approaches to satisfy
    capability or product objectives, indentify
    technology necessary to satisfy objectives,
    and assign criticality

    objectives
    weighted   coordination
    time–frames

3. Technology Team (first activity)

    Assemble technologies by common headings
    and levels of difficulty

4. Technology Team (second activity)

    Determine resources required for technology
    by date stipulated by interdisciplinary teams

5. Fixed Budget

6. Computer Program

7. Balanced Exploratory Development Funds Allocation

FIGURE 3. Torque flow chart. *From Nutt [9, p. 245].*

able. The simulation model (objective function) is

$$U = \sum_{j=1}^{N} (C_j)(W_j)(Cf)(t_j) \tag{8-8}$$

where

U = utility of the particular technological state-of-the-art (SOA) advance required to realize the system/subsystem options

N = number of systems supported by the proposed SOA advance

C_j = criticality of the SOA advance to the jth system supported

W_j = relative normalized weight or importance of the desired objective supported by the jth system

Cf = the ratio of funds allocated to the technology package to which this SOA advance belongs in the first year to the total funds required to achieve this technology completely

t_j = timeliness function

Where the SOA advance is completed between the earliest and latest dates, t_j is equal to 1.0 and the jth system can use the advance. Where the SOA advance is completed more than 2 years before the earliest date E or later than 2 years after the latest date L, $t = 0$. Intermediate values between zero and 1.0 are determined by the symmetrical trapezoidal shape of the timeliness function (e.g., Figure 4).

The computer, using the budget allocation, selects from the combination of allocation levels for the various technologies the set that provides the maximum utility for the funds available. The outputs are then formatted by organization, project, and task, showing funds allocated. Other output formats can be devised that show the data in various forms (e.g., priority lists) depending on the requirements of the user.

Certainly there is a vital need for developing and using methods and techniques that show more clearly the relevance and contribution of funds spent on research and initial development to future operational products, processes, or capabilities. TORQUE is one management science tool/ technique that aids decision making in this area.

G. Updated PERT

The update of classical PERT [10] resulted from an experiment that set out to improve the procedure for estimating the mean and variance of the hypothetical distribution of performance times. The underlying assumption of the experimenters was that estimation of the duration of some new activity is based upon one's experience; that is, estimators mentally divide an activity into subactivities in which they have had direct experience. The experience data given to the subjects were random samples of a beta distribution whose parameters were known only to the experimenter.

An analysis of variance on the estimates of the mean indicated that CPM estimates were significantly different (2.5% level) from the PERT methods based on percentiles and, furthermore, that there were no significant differences among the percentile methods ($d = $ 0th, 5th, or 10th percentiles). The indication is that the PERT percentile estimates are

FIGURE 4. The timeliness function.

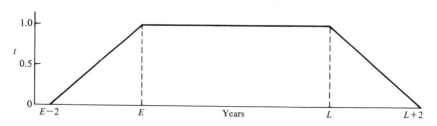

unbiased, whereas the CPM estimates are biased on the high side by approximately 4%.

In view of the statistical soundness of the estimates of variance based on the 5th and 95th percentiles to variations in the shape of performance time distribution, and on the basis of the experimental results, the following formulas are recommended for estimating the mean μ and variance σ^2 of the hypothetical distribution of activity duration times:

$$\mu = (a_5 + 4m + b_{95})/6 \tag{8-9}$$

$$\sigma^2 = (b_{95} - a_5)^2/10.2 \tag{8-10}$$

where

$a_5 = $ optimistic performance time at the 5th percentile
$m = $ mode of the distribution
$b_{95} = $ pessimistic performance time at the 95th percentile

H. Analysis of Uncertainty Model

This model [11] is a method for analyzing the resolution of uncertainty over time associated with the introduction of a new product/process. The analysis is undertaken within a capital budgeting framework that yields a Go or No-go decision for the product/process under consideration. The technique is applicable to the R&D problem/decision areas of decision making, strategy, economic analysis, research and development budgeting, and project selection.

Conventionally, the firm considering a new product or process uses traditional cash-flow engineering economy principles to aid in decision making and budgeting. Unfortunately, this method does not consider the fact that the greater portion of risk occurs and is resolved early in the life of the product. This model develops a value for the measure of risk that may be graphically portrayed versus time to allow a determination of the firm's risk posture in view of acceptance of the proposed product (or its rejection).

The expected value of present worth (PW) at time zero is

$$PW = \sum_{t=0}^{n} \frac{\bar{A}_t}{(1 + i)^t} \tag{8-11}$$

where $\bar{A}_t$ is the expected value of net cash flow in period t and i is the risk-free interest rate.

As a measure of absolute risk, the standard deviation of the probability distribution of PW is determined by

$$S_0 = \left[\sum_x PW_x^2 P_x - (\overline{PW})^2 \right]^{1/2} \tag{8-12}$$

where S_0 indicates standard error for time zero, PW_x is the present worth for series x of net cash flows, covering all periods (periods refers to sections of a tabular probability of occurrence/net cash flow layout projected over time considering conditional events), and P_x is the probability of occurrence of that series.

In order to measure the expected resolution over time (CV_t), a statistic is constructed to approximate relative uncertainty at a moment in time. This statistic is the ratio

$$CV_t = S_t/\overline{PW} \qquad (8\text{-}13)$$

where S_t represents the average standard error of the various branches of the probability tree that resulted from the tabular layout of probability/net cash flow, and is computed in the following manner:

1. Calculate present worth of all expected cash flows. Label these present values Y_{gt} (gth PW in period t).
2. Determine the total node value T for each node and tip of the probability tree by the following method.
   a. For each node, compute the expected value of all future Y_{gt} in that branch.
   b. For each node and each branch tip, sum the Y_{gt} involved in reaching that node or tip from time zero.
   c. Add (a) and (b) to obtain T.
3. Compute the weighted sum of squares of the total node values for each period by

$$V_t = \sum_h T_{ht}^2 \cdot P_{ht} \qquad (8\text{-}14)$$

where T_{ht} is the hth total node value at the end of period t and P_{ht} is the probability of reaching that node or branch tip. When $t = 0$, there is but one total node value—the expected value of net present worth of the product.

4. Determine s_t for each period by

$$S_t = (V_n - V_t)^{1/2} \qquad (8\text{-}15)$$

where n is the last period in the probability tree.

CV_t now may be plotted against time to portray the uncertainty resolution trend of the particular product. Management's decision is then based on conventional expected value results tempered by the uncertainty trend.

The actual concern will be with the marginal impact of the new product on the resolution of uncertainty for the firm's entire product mix; i.e., account must be taken of the effect of the new product on the cash flows of existing products. To determine the risk of the product combination, the standard deviation of the probability distribution of possible total node values for a combination of m products at the end of period t is calculated by

$$\sigma_t = \left(\sum_{j=1}^{m} \sum_{k=1}^{m} \sigma_{jkt} \right)^{1/2} \qquad (8\text{-}16)$$

where σ_{jkt} is the covariance at time t between possible total node values of products j and k,

$$\sigma_{jkt} = r_{jkt} S_{jt} S_{kt} \qquad (8\text{-}17)$$

where r_{jkt} is the expected correlation between possible total node values for products j and k. The four-step calculation of S_t is as previously discussed. When $j = k$, σ_{jkt} becomes S_{jt}^2. The degree of correlation r is estimated by management on the basis of past experience and future forecasts.

The incremental risk, IR, of taking on a new product is determined by subtracting the variance at time zero for the combination of existing products from the variance for the combination of existing products, σ_E^2, plus the new product, σ_N^2, and taking the square root of the difference, that is,

$$IR = [\sigma_0^2 - (\sigma_E^2 + \sigma_N^2)]^{1/2} \qquad (8\text{-}18)$$

The graphical portrayal may be presented by plotting the risk profile for the combination of existing products plus the new product against that for the combination of existing products alone.

The Go/No-go decision for each product may be made either on an individual basis or according to some overall set of constraints. Suggested constraints might be a profitability index such as

$$PI = (IPW + PWO)/PWO \qquad (8\text{-}19)$$

where IPW is the incremental expected value of net present worth of the product and PWO is the expected value of present worth of cash outflows for the product. Certain confidence intervals may be established; for example, the profitability index should not be less than one, or incremental CV_t should not exceed certain stated values. If the product or project under consideration satisfies these constraints, it should be undertaken; otherwise, it should be rejected.

By virtue of this management science tool/technique, management is able to make more rational new-product decisions. Certainly, this information about the resolution of uncertainty is valuable in planning for new products and in balancing the risk of the firm over time.

I. A Model for Technical Problem Solving

This model [12] is not a quantitative method but merely presents a logical flow chart of an information-processing theory of human problem solving as applied to the initial technical feasibility studies for an R&D project. Since the technique is a guide for the attack on a technical problem, it can be applied almost across the board, with great capability in the areas of decision making, planning, project selection, product design, product development, innovation, and certainly to organizational effectiveness.

The ability to understand the decision behavior of an individual engaged in the process of solving a technical problem will enable management to improve the organizational support of the process and thereby improve the effectiveness of the total R&D effort.

A technical problem often has no correct, or even best, solution (in the long run). In fact there is frequently no terminal state: Both problems and solutions are dynamic. The dynamic nature of problems and solutions is included in the model as a procedure in which the heuristics employed by human problem solvers are first identified and then described in a programmable model.

Three major sources of initial information—the customer, the environment, and experience—are the input values. The technical problem-solving process is then divided into two streams that are initiated and pursued simultaneously. The first set of processes (10–50 in the model flowchart, Figure 5) involves the generation of critical dimensions against which the solution is to be judged. Of course, the customer's requirements are the foundation of the whole process. Critical dimensions may be dictated explicitly by the customer or management or implicitly by management policy.

The distinction between fixed requirements and the remaining dimensions can best be described when the technical quality of a solution is considered as the composite score of its evaluation on each dimension. For fixed requirements, the score on each dimension is measured on a discontinuous scale. On the remaining critical dimensions, a solution is scored on a linear scale over an acceptable range of the dimension. Measurable fixed requirements and the remaining critical dimensions are then ranked in order of importance (50). The resulting list of fixed requirements and critical dimensions are information inputs to the second part of the process.

The generation of alternative solutions (60) is similar to the generation of critical dimensions, although the use of information sources is more evenly distributed. Based on the information used to generate the approaches, alternatives are first ranked on the probability that they will be acceptable on all critical dimensions (70). Following this, all further consideration is limited to a subset of n top-ranking alternatives (80). The

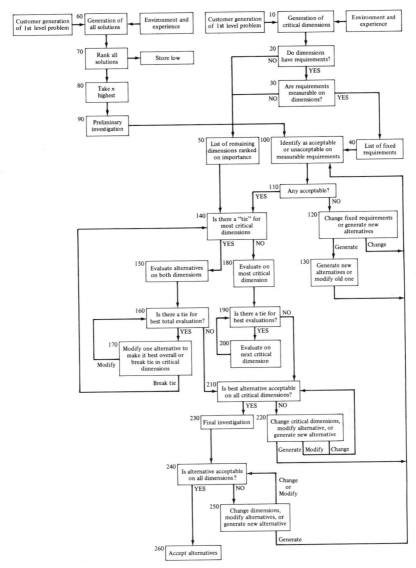

FIGURE 5. A model for technical problem solving. *From Frischmuth and Allen* *[12, p. 60].*

problem solver now subjects the remaining approaches to a preliminary investigation (90) in which literature and vendors are consulted and simple analyses performed. On the basis of this investigation, and of the measurable fixed requirements (40), the approaches are identified as acceptable or unacceptable (100). If no approaches meet the requirements (110, No),

TABLE 2. Management Science Tools and Techniques Applied to R&D Problem/Decision Areas

Management science tools and techniques	R&D problem/decision area														
	Decision making	Strategy	Planning, forecasting, and prediction	Economic analysis	Optimization and resource allocation	Cost estimation	R&D budgeting	R&D evaluation	Project selection	Product design	Product/process development	Product/process evaluation	Project scheduling and control	Innovation	Organizational effectiveness
Real-time computer	×				×							×	×		×
Dynostat	×	×		×	×		×			×					
Research projects selection model	×	×	×						×						
Delphi procedure									×						
CONTEST	×					×	×	×	×			×			
TORQUE	×								×						
Updated PERT		×					×				×		×		
Analysis of uncertainty model	×		×	×					×						
Technical problem-solving model	×								×	×	×			×	×

the main stream of events is interrupted. The problem solver must now attempt to change the fixed requirements or generate new alternatives.

Those alternatives that are judged acceptable (110, Yes) now enter an evaluation process (140). If, on the list of critical dimensions, there are two that are equally important, the alternatives are evaluated on both (150). Ties on overall evaluation (160, Yes) at this point are evaluated on the third-ranked dimension (200). To break the tie in total evaluation, the problem solver may modify one alternative such that its overall evaluation is higher. In this case, the process proceeds through steps 160–210. If two approaches tie for best evaluation (190, Yes), they are evaluated on the next most critical dimension and the overall evaluations are compared. This loop is continued until the tie is broken (190, No) and the process continues to step 210.

The remaining favored alternative is evaluated on all critical dimensions (210) and judged acceptable or unacceptable. If the approach is not acceptable, the problem solver repeats processes 120 and 130 but changes critical dimensions rather than fixed requirements in step 220 and returns to 210. When the alternative is acceptable on all critical dimensions, it is subjected to the final investigation (230). Based on the results, the alternative is evaluated on all dimensions. If it is not acceptable, the process is repeated on the combination of processes 120, 130, and 220 in step 250. When the alternative solution is acceptable on all dimensions, the process moves on to the next level of the problem.

This model serves as an excellent guide for the management scientist/engineer in technical problem solving. It is also useful for management because it allows some insight into the process by which a particular solution or suggestion originated. The concept of a variable problem, as addressed in the model, is realistic. It demonstrates that the problem-solving process is a blend of two extremes: the fixed-answer problem and the generated problem. The combination of these two processes is important in the light of its impact upon the problem-solving process.

Table 2 summarizes the applicability of the various tools and techniques discussed in this section to R&D problem/decision areas.

REVIEW QUESTIONS

1. Define basic research.

2. Briefly explain applied research.

3. Discuss the role of the research and development effort in an organization and list some specific R&D objectives that should be considered.

4. The basic nature of research and development requires high costs and long-term investments. Do you agree or disagree with this statement? Why?

5. Briefly explain how dynamic programming can be used in research and development problems.

6. In what way can input–output analysis be useful to the management of the development program?

7. Discuss marginal utility and expected utility. Explain the differences between them.

8. How do you compute the expected value of a development project? Give an example.

9. Describe the interrelated stages involved in the application of industrial dynamics to R&D problems.

10. Cost and reliability are two important factors in value analysis. Which would you say is more important for this system?

11. What are the cost coefficients of the Dynostat objective function?

12. Describe the Delphi process. Do you think this method is just an averaging of view points? Explain.

13. Discuss the phase 1 and phase 2 processes of CONTEST.

14. List four questions that should be answered by TORQUE and explain how this technique supplies the answers.

15. Explain the basic concept of updated PERT.

REFERENCES

1. National Science Foundation, Basic Research, Applied Research, and Development in Industry, *Surveys of Science Resources Series*. Washington, D.C.: NSF 67-12, 1965.
2. Villers, R., *Research and Development—Planning and Control*. New York: Financial Executives Research Foundation, 1964.
3. Mundel, M. E., *A Conceptual Framework for the Management Scientist*. New York: McGraw-Hill, 1967.
4. Ferguson, R. L., and Jones, C. H., A Computer Aided Decision System, *Management Science* 15, 550–557 (June 1969).
5. Gibson, J. A., and Coombes, G. E., A Parallel Optimum Seeking Technique— Dynostat, *IEEE Transactions on Systems Science and Cybernetics* 55C-6, 197–208 (July 1970).
6. Atkinson, A. C., and Bobis, A. H., A Mathematical Basis for the Selection of Research Projects, *IEEE Transactions on Engineering Management* EM-16, 2–8 (February 1969).
7. Martino, J. P., An Experiment with the Delphi Procedure for Long-Range Forecasting, *IEEE Transactions on Engineering Management* EM-15, 138–144 (September 1968).
8. Skolnick, A., A Structure and Scoring Method for Judging Alternatives, *IEEE Transactions on Engineering Management* EM-16, 72–83 (May 1969).
9. Nutt, A. B., Testing TORQUE—A Quantitative R&D Resource-Allocation

System, *IEEE Transactions on Engineering Management* EM-16, 243–248 (November 1969).
10. Moder, J. J., and Rodgers, E, G., Judgment Estimates of the Moments of PERT Type Distribution, *Management Science* 15, B76–82 (October 1968).
11. Van Horne, J. C., The Analysis of Uncertainty Resolution in Capital Budgeting for New Products, *Management Science* 15, B376–386 (April 1969).
12. Frischmuth, D. S., and Allen, T. J., A Model for the Description and Evaluation of Technical Problem Solving, *IEEE Transaction on Engineering Management* EM-16, 58–64 (May 1969).

RECOMMENDATIONS FOR FURTHER READING

Barry, Christopher B., and Wildt, Albert R., Statistical Model Comparison in Marketing Research, *Management Science* 24 (4), 372–392 (December 1977).

Bass, L. M., The Management of Technical Programs, in *Praeger Special Studies in International Economics and Development*. New York: Praeger, 1965.

Boot, J. C., *Mathematical Reasoning in Economics and Management Science*. Englewood Cliffs, N. J.: Prentice-Hall, 1967.

Dean, B. V., *Operations Research in Research and Development*. New York: Wiley, 1963.

Ein Dor, Phillip, and Segev, Eli, Strategic Planning for Management Information Systems, *Management Science* 24 (15), 1631–1641 (November 1978).

Glover, F., Hultz, J., Klingmand, D., et al., Generalized Networks: A Fundamental Computer-Based Planning Tool, *Management Science* 24 (12), 1209–1220 (August 1978).

Gregg, J. V., *Mathematical Trend Curves and Aid to Forecasting*. London: Oliver & Boyd, 1964.

Karger, D. W., and Murdick, R. G., *Managing Engineering and Research*. New York: Industrial Press, 1969.

Liberman, Varda, and Yechiali, Uri, On the Hotel Overbooking Problem—An Inventory System with Stochastic Cancellations, *Management Science* 24 (11), 1117–1126 (July 1978).

Moundalexis, J., and Lichtenberg, W., Input/Output Analysis of Organizations Having Intangible Outputs: 10A010 Technique, *Proceedings AIIE* (1966).

Roberts, E. B., *The Dynamics of Research and Development*. New York: Harper & Row, 1964.

Roman, D. D., *Research and Development Management: The Economics and Administration of Technology*. New York: Appleton-Century-Crofts, 1968.

Schendel, Dan, and Patton, G. Richard, A Simultaneous Equation Model of Corporate Strategy, *Management Science* 24 (15), 1611–1621 (November 1978).

Schwartz, S. L., and Vertinsky, I., Multi-Attribute Investment Decisions: A Study of R&D Project Selection, *Management Science* 24 (3), 285–301 (November 1977).

Weinberg, Charles B., Jointly Optimal Sales Commissions for Nonincome Maximizing Sales Forces, *Management Science* 24 (12), 1252–1258 (August 1978).

Weinberg, Charles B., and Shachmut, Kenneth M., ARTS Plan: A Model Based System for Use in Planning a Performing Arts Series, *Management Science* 24 (6), 654–665 (February 1978).

PART III
MANAGEMENT SCIENCE PROBLEMS
AND PROMISES

Chapter 9 discusses the present status of management science, its achievements and problems of application. During the last decade, emphasis was placed on the development of management science tools and techniques. Today, effort is being concentrated on applying the previously developed tools and techniques to actual problems.

The scientific approach has been demonstrated to be a successful approach to problem solving. The basic tools and techniques are available. Computational capability exists via the computer. However, there are problems in the process of utilizing management science tools and techniques.

This chapter discusses these problems and explores ways and means of overcoming these problems. The principal emphasis is for management science to become more problem-oriented and solve real life problems.

Chapter 9
Management Science Problems and Promises

INTRODUCTION

Management science today finds itself in a rather awkward position. During the last two decades there has been an unprecedented development of new managerial, engineering, and mathematical techniques for providing improved operating effectiveness to business and governmental operations. Yet many engineers, managers, and scholars question whether these new developments have actually provided significant improvements in dealing with the operations management problems of the current time, as measured by actual operating results. The essence of this position is that the business results of profitability and efficiency still do not demonstrate the effectiveness of these techniques in any generally accepted way. The problem appears to be that management scientists have given far more attention to the development and communication of the analytical bases of these techniques than to the systemmatic ways in which the techniques can be put to widespread organizational use. The introduction of the methodologies of quantitative analysis, automation, information processing, control engineering, and management science has the potential of getting to the roots of business and governmental operations. However, to produce positive results from these techniques, their introduction must be accompanied by the creation of an equally powerful managerial and engineering decision-making and operating framework to put the techniques to work. Most firms are far more ready to install a management science activity than to support it with the organizational modifications suggested by the analytical work. Before management science can become an effective tool it will be necessary to remove the numerous decision-making obstacles that currently inhibit any real results from these techniques, and the management science techniques themselves must be reoriented to emphasize the institutional, social, and political aspects of the work as well as the technical side. Technical development simply is not enough for such strong major new areas of analytical methodology that management science proposes. To make this methodology useful, there must also be the development of frameworks that are effective enough to handle the important decisions involved in the practical introduction and maintenance of the methodology.

Because the future of management science depends on understanding the problems of applying its tools and techniques and on finding solutions for these problems, it is important to discuss these problems and to suggest future approaches to management science applications.

I. PROBLEM AREAS

A. Management Science Is a Young Profession

When the top executives find themselves worried because they feel outdated by the computer, they hire young management scientists, fresh from college, in order to give the company a modern approach. Those young management scientists then promise the company a great information system that will solve all management's problems. Before enough years go by to develop the information system, one finds that the management scientists begin to leave the company, some because of the realization of the enormity of their tasks, others because rival companies are willing to pay more for their experience. Eventually, management gets discouraged and the promised system dies.

B. Approach May Not Be Realistic

It is felt by some that one of the major problems dealing with the acceptance of management science arises as a result of enthusiastic young practitioners fresh from a college and their professors who have had no experience with real, in-context management problems. Consequently, these new practitioners offer mathematical solutions to management that management considers unrealistic and out-of-context. The argument is not diminished by the fact that many professors in management science have consulting practices on the side, for, as surgeons bury their failures, consultants walk away from theirs. Because of this situation, the professors turn out suggestions, articles, and mathematical models in the hope of solving the problems they pose to themselves, based on assumptions they find they need to make but which are untenable in just about every management situation. These elegant models based on untenable assumptions convince many a manager to forego the installation of the management scientist.

C. Management Is Both Art and Science

The success of a business is dependent on profits, and profits are largely dependent on decisions concerning future events. A management system can assist in these decisions by a good and timely analysis of what would be the outcome of making such a decision if the assumptions based on historical happenings continue, but the basic ingredients of a future decision are judgment, initiative, and common sense, elements extensively lacking in an automated management system. Yet many freshly educated management scientists promote such a concept. It is common for the practitioner of this new science to feel that the executive of the future will make all the decisions by computer, so that the company's operations should be restructured to exploit the potential of the computer. It is also common for the management science purveyor to want to revolutionize the company with a fully integrated corporate-wide system, to utilize an automated management system to answer any management question almost before it is thought of by the manager. And all these concepts generally lead to problems in obtaining a valid management information system.

D. The Human Element of Systems Design

Systems scientists are often accused of a simple lack of feeling for people and an inability to go beyond a disastrous "machine concept" of humanity. It is common today to find those who feel that technology has dehumanized people and that technology is the root cause of all social polarization and environmental problems. To those people, often found in many of the social science fields, technology in the hands of the management scientist is reducing humans to cogs in the productive process. Some people fear technology because they think that it will replace them; others criticize the systems approach as a result of personal experiences in which the management scientist developed a disastrous, short-term solution to a complex, long-range problem.

It is also said that management scientists are insensitive toward people, that the idea behind scientific management is that of the smooth-running machine powered by economically motivated workers who are comparable to interchangeable units. The accuser of the management scientist claims that the management scientist does not question cold organizational objectives but sacrifices others in an attempt to meet these objectives.

II. APPROACHES TO SOLUTIONS

A. Middle-Ground Approach

There are two extreme views of management science: there is the glowingly optimistic can-do-everything concept held by some, and there is the "It doesn't work, it's too unrealistic" view held by others. A realistic middle-ground approach would seem to be necessary for management science if it is to develop a strong growth role in the future. In the past, management science was hindered by an inability to do the computations required of its approaches. Today, the computer provides the mathematical solutions, but the management scientist must be able to incorporate physical, emotional, psychological, and organizational aspects, as well as the purely management science ones, before these solutions can be meaningful to managerial decision making. Today's management scientist is capable of using the computer for arriving at solutions to specific problems framed in specific assumptions. Because of this, the problems that are most amenable to the management science approach are those problems that are not usually encountered on a routine basis. It is the nonrepetitive aspect that makes computational efficiency less critical and allows minimal attention to be devoted to planning data input and output formats. At the same time, this nonroutine requirement sometimes yields such information that it becomes a formalized method of thought about the problem and thus helps reduce management risk in more commonly occurring problems. But the management scientist must be willing and able to realize the inherent shortcomings of the model so that management can evaluate the information with as little bias as possible. This approach will result in an increased trust in the management science information. But it is also important to be aware of the potential psychological consequences inherent in the installation of a management science, management information system.

B. User Participation

Before setting up a management information system, one must investigate the nature of the overall organizational climate for its potential effect on the information system. Critical aspects are management and employee attitudes, organizational cohesiveness, and organizational culture. The attitude of top management should be one of full support for the approach. Lines of communication should be open among all different levels of management and employees and between line and staff. An overall feeling of trust and cooperative spirit should exist, free of excessive concern over job stability and security. If any significant problems exist in these areas, the management scientist will have to create a better atmosphere before any newly developed system will obtain its goals.

Participation, especially from the operating managers, in decisions pertaining to the basic structure of the system should be on a widespread basis. In addition, every attempt should be made to achieve harmony between individual and organizational goals.

It is essential for the management scientist to create attainable goals for employees involved in the system being developed if the employees are going to accept and learn to use the system. At the same time it is necessary to make the purpose and characteristics of the system as clear as possible to those who have any direct contact with it. Effective use or support cannot be obtained if there is confusion or lack of understanding. It is also important to minimize system errors in the early stages of program initiation. Numerous errors during the first few months of operation can jeopardize a potentially good system by undermining the users' faith and trust in the output information. In the overall interests of the organization, and for the most effective system, it may even be necessary to have certain tasks that are compatible with computer processing remain in the hands of individuals so that they may retain their sense of dignity and importance along with the feeling of being worthy contributors to the needs of the organization. Acceptance of the system is not likely if the system usurps these individual needs. There must also be a shift away from awards based on individual accomplishments to rewards based on group achievement. Often the main resistance of managers to systems programs results from their fear of losing control over their subunits. Finally the systems designer should be careful not to overwhelm the user with a large volume of output which the user either cannot understand or cannot use effectively, while providing the user with the necessary accurate output.

C. People-Oriented Systems

Although management scientists are technologists, they must also have the capability to work with people if the system is to work within the organization. Without such a coordinated approach the myths about and antagonisms toward management science could easily prevent its effective implementation.

In engineering and managing the creation of a system there are two significant problems. The first problem has to do with the major technological complexities involved in establishing the system. The second problem has to do with the massive human, organizational, and timing complexities that are involved. These frameworks will have an effect on the job assignments of people and will alter the managerial approach to the operation of the enterprise.

The need is for much tighter integration among human actions and machine and information flow. This integration of personnel, material, data, and equipment indicates a specialized type of people–machine–information system. There is much technology now available to develop a coordinated people–machine–information operating framework that can serve as the basis for truly effective enterprise management in the major areas of the firm. However, the management scientist is still not capable of handling the human element in the technology, and it is the human element and organizational complexities that are the major obstacles to the development of effective operating frameworks. Today's approach toward this operating framework, neglecting the human element, predicts a future organization as a total systems organization in which the computer will dominate the information system. This future organization will increase the emphasis on precision planning and more creativity in decision making, necessitating the reduction of middle management to clerk status and causing the top management function to be divided among several people who will act in concert as the president.

Such a concept may indeed be the logical organizational framework of the future, but this approach neglects the vital human element. Before such a system can become a reality, it will be necessary for the social sciences to develop sufficiently; otherwise it is likely that extensive human resistance will prevent the system from taking root. The future of management science appears to depend more on a gradual blending of system and staff. Without due recognition of the human element it will be found that the disillusionment with management science will prevent its general application.

D. Problem-Solving Approach

In evaluating management science as a technical discipline we notice that focus is being placed on different aspects of management science depending on the school. For the purpose of discussion, these aspects may be classified as

1. Mathematical- and technique-oriented
2. Organization-theory- and human-behavior-oriented
3. Problem-oriented.

The mathematical- and technique-oriented schools place primary emphasis on existing model-building and model-solving techniques.

The organization-theory- and human-behavior-oriented schools focus their attention on the "people problem." This aspect is emphasized primarily in the schools of business, public administration, and behavioral sciences.

The problem-oriented schools place emphasis on management problems and their applications. Technique classification is replaced by problem classification. Emphasis is placed on not finding a problem for a technique, but finding the perfect technique for a nonroutine problem. The use of approximation methods and heuristics is stressed.

Heuristic reasoning has been described as reasoning, regarded as neither final nor strict, whose purpose is to solve the present problem. It makes no claim for the best properties for a solution but evaluates a solution as an improvement over previously available ones. Naturally the training of people in the problem-oriented approach is considerably different and probably more difficult than the normal teaching of techniques only.

In the previous discussion it has been pointed out that for management science to prosper, emphasis must be on problem solution rather than technique development. However, this does not mean that technique development should cease. On the contrary, problem solving requires the use of techniques. Therefore, the management science discipline should provide a solid foundation in mathematics and in tools and techniques, with the emphasis on problem solving. It is this overall coordinated approach, impossible to teach in the university, that will allow management science to grow in the future.

III. FUTURE OF MANAGEMENT SCIENCE

It is true that the future cannot be predicted with certainty, but if we consider the past there are some trends that can be projected.

As it has been shown from experience, the demand for management science tools and techniques will increase in the future. Management science will be used in more fields than ever before, including social subjects, environmental, health, public planning, and transportation. More nonmathematical constraints in areas such as political restrictions in public planning can be quantified and incorporated into mathematical models.

The continued growth of the management science field is evidenced by the fact that we now find in many of the larger companies the operations research and computer functions under a management science staff organization reporting to the vice-presidential level. The trend is for the management science functions to be situated as a high-level staff organization with easy access to top management as well as the line organizations.

More college graduates with management science degrees will assume managerial positions, resulting in a higher demand for the problem-solving tools and techniques of management science.

Although there is a need for advancement of the mathematical techniques and there will be much progress in this field, there is a greater need for development of lower-cost utilization of computers. On the other hand, the recent trends in computer technology allow users to be more productive. Today's computers provide improved service and allow the users to expand the time-sharing and on-line functions. The capacity of new computers to accommodate a huge size data base, provides up-to-date information instantly through on-line terminals. This information is the essential tool of the decision-making process and will result in the application of advanced mathematical models to the complex real world problems, providing accurate and reliable solutions.

Total systems applications of management science require timely, complete, and accurate data. In a total systems approach, the knowledge of what data are required may prevent management science from meeting these three requirements. Thus it appears that many management information systems may be temporarily limited to a divisional or group-operating level until there is a significant increase in communication ability throughout most of the world. But even with these limitations being imposed on management science, it is apparent that eventually all companies will find a need for computer-based management information systems. The systems will not be as complete or infalliable as many of their advocates claim, but at the same time management science techniques will remain essential in giving management an inside look at the "what if" questions. In addition, the manager will still be capable of judgment and intuition, for the availability of every piece of information about one's own company is not necessarily as significant as having just one right piece of outside data.

In brief, management science has a promising future. The scientific approach has been demonstrated as a good approach to problem solving. The basic tools and techniques are available. Good computational capability exists via the computer. All that is required is for management science to become more problem-oriented and solve real world problems.

RECOMMENDATIONS FOR FURTHER READING

Textbooks

Cook, T. A., and Russell, R. A., *Introduction to Management Science.* Englewood Cliffs, New Jersey: Prentice-Hall, 1977.

Eppen, Gary D., and Gould, F. J., *Quantitative Concepts for Management.* Englewood Cliffs, New Jersey: Prentice-Hall, 1979.

Johnson, Rodney D., and Siskin, Bernard R., *Quantitative Techniques for Business Decisions.* Englewood Cliffs, New Jersey: Prentice-Hall, 1976.

Levin, Richard I., and Kirkpatrick, Charles A., *Quantitative Approaches to Management.* New York: McGraw-Hill, 3rd ed., 1975.

Shore, Barry, *Quantitative Methods for Business Decisions—Text and Cases.* New York: McGraw-Hill, 1978.

Articles

Adam, Everett E., Jr., Priority Assignment of OSHA Safety Inspectors, *Management Science* 24 (15), 1642–1649 (November 1978).

Blumstein, Alfred, Management Science to Aid the Manager—An Example from the Criminal Justice System, *Sloan Management Review* 15 (1), 35–48 (Fall 1973).

Bodily, Samuel E., Police Sector Design Incorporating Preferences of Interest Groups for Equality and Efficiency, *Management Science* 24 (12), 1301–1313 (August 1978).

Braustein, Daniel N., Psychologists and Management Theory—Getting Them All Together, *Interfaces (TIMS)* 6 (1), 50–52 (November 1975).

Bunn, D. W., and Mustafaoglu, M. M., Forecasting Political Risk, *Management Science* 24 (15), 1557–1567 (November 1978).

Carbone, Robert, and Longini, Richard L., A Feedback Model for Automated Real Estate Assessment, *Management Science* 24 (3), 241–248 (November 1977).

Cargill, Thomas F., and Eadington, William R., Nevada's Gaming Revenues: Time Characteristics and Forecasting, *Management Science* 24 (12), 1221–1230 (August 1978).

Debanne, J. G., Management Science in Energy Policy—A Case History and Success Story, *Interfaces* 5 (2), 1–21 (February 1975).

Eck, Roger D., Treading Softly with Management Science, *Arizona Business* 22 (1), 3–7 (January 1975).

Gapinski, James H., and Tuckman, Howard P., Amtrak, Auto-Train, and Vacation Travel to Florida: Little Trains That Could, *Management Science* 24 (11).

Ginzberg, Michael J., Steps Toward More Effective Implementation of MS and MIS, *Interfaces* 8 (3), 57–63 (May 1978).

Graham, Robert J., Problem and Opportunity Identification in Management Science, *Interfaces (TIMS)* 6 (4), 79–82 (August 1976).

Graham, Robert J., The First Step to Successful Implementation of Management Science, *Columbia Journal of World Business* 12 (3), 66–72 (Fall 1977).

Harris, Carl M., and Moitra, Soumyo D., On the Transfer of Some OR/MS Technology to Criminal Justice, *Interfaces (TIMS)* 9 (1), 78–86 (November 1978).

Hopkins, David S. P., Computer Models Employed in University Administration: The Stanford Experience, *Interfaces (TIMS)* 9 (2), 13–22 (February 1979).

Lee, Thomas D., Graves, Robert J., and McGinnis, Leon F., A Procedure for Air Monitoring Instrumentation Location, *Management Science* 24 (14), 1451–1461 (October 1978).

McNally, James P., How Management Science Can Save You Money, *Price Waterhouse Review* 22 (2), 18–23 (1977).

Nackel, John G., Goldman, Jay, and Fairman, William L., A Group Decision Process for Resource Allocation in the Health Setting, *Management Science* 24 (12), 1259–1267 (August 1978).

Noble, Carl E., Why Wait for Crises to Reveal the Unusual Opportunities for Management Science Applications in Your Organization, *Interfaces (TIMS)* 7 (4), 50–55 (August 1977).

Shapiro, Alan C., Incentive Systems and the Implementation of Management Science—A Spare Parts Application, *Interfaces (TIMS)* 7 (1), 14–17 (November 1976).

Stradtherr, Mark A., and Rudd, Dale F., Resource Management in the Petrochemical Industry, *Management Science* 24 (7), 740–746 (March 1978).

Shycon, Harvey N., Perspectives on MS Applications, *Interfaces* 4 (4), 39–42 (August 1974).

Shycon, Harvey N., Perspectives on MS Applications, *Interfaces (TIMS)* 6 (4), 27–30 (August 1976).

Vicino, Franco L., and Bass, Bernard M., Lifespace Variables and Managerial Success, *Journal of Applied Psychology* 63 (1), 81–88 (February 1978).

PART IV
APPENDIXES

Appendix 1
Computer Software for Accounting and Financial Analysis

INTRODUCTION

In the context of accounting and financial analysis, the large-system user generally has an easier choice of alternative software, since such systems, being in the vanguard, have been much researched and discussed in the literature. Hardware manufacturers have extensive compatible program libraries available for these large systems, and the users themselves have sophisticated EDP capability, with trained personnel able to develop software tailored to specific needs.

The smaller user, however, is faced with a vast array of choices, which can be condensed into four major groups.

1. Manual systems
2. Service bureaus
3. Time-sharing systems
4. Minicomputers

The alternative selected will depend upon several factors, the more significant being economic viability, the size of the firm, the types of application, and the computer expertise available to the firm.

I. SERVICE BUREAUS

In general, those outfits choosing this arrangement have a tendency to use the service for common high-volume accounting data such as accounts receivable, accounts payable, billing, invoicing, inventory, and sales analysis. Automating functions unique to the user, like production planning and scheduling, can be costly; however, for a small user this can still be more economical than having to face start-up costs of system installation.

In a study of service bureau use [1] only 13% of the respondents reported direct savings in clerical costs. However, this ignored the question of efficiency, that is, whether the users' previous systems had been as efficient as those of the service bureau. The increase in both speed and accuracy may compensate for higher bureau costs. Also, more and more users are beginning to go beyond routine applications to realize increased benefits.

Problems with service bureaus revolve around slow turnaround time—often a minimum of a week. Errors necessitate both additional cost and more turnaround time. Advantages, on the other hand, primarily involve the lack of initial capital investment in both equipment and personnel.

II. TIME SHARING AND MINICOMPUTERS

Time sharing is particularly well suited for applications where the computational work is complex but input–output requirements are minimal. This is because the commonly used input–output device—the typewriter terminal—cannot easily handle a large volume of input and output. However, with high-speed printing services and various inexpensive off-line storage devices, such as tape cassettes, high-volume input and output applications, have become economically feasible. The major advantage of time sharing lies in the immediate feedback of an on-line system. A job does not have to be set aside for a couple of days while being processed, as it would if a service bureau were being used.

Minicomputers are often ideally suited for local CPA firms that spend a large percentage of time in general ledger, financial, and tax statement processing for their write-up clients. Often the firm employs a large clerical staff to post the books and typists to type out the reports; a small accounting computer can substantially cut costs. The basic configuration consists of a small central processing unit and keyboard input–output; often a magnetic- or paper-tape reader and an optional high-speed printer are available. Usually manufacturer-designed software, with some editing flexibility, can also be obtained. It is safe to say that the target users for which such systems are designed would probably find it difficult to develop their own software; however, many systems are programmable in the more common languages, so the possibility of software development is there.

Major hardware considerations involve the matching of system capability and the user's requirements. Important factors include:

Available software and its cost

Maintenance services provided by the vendor and their cost

Obsolescence and the efforts the vendor is willing to make to upgrade the system

How far the system can be expanded without necessitating fundamental changes

Training of user personnel and the vendor's training services

Ease of system use.

A. Time-Sharing Software

There are over 100 independent time-sharing organizations with services ranging from mere provision of computer time to sophisticated package programs. Commonly used vendors include Compu-Serv, Tymshare, Comshare, and the General Electric Company. The American Institute of Certified Public Accountants (AICPA) has established common libraries for the exchange of time-sharing programs among CPAs and is also looking for contributory practitioners.

The AICPA general library is available to members on Comshare's system [2]. The general library contains programs for the following areas:

Accounting and auditing

Financial analysis

Investment analysis

Investment plan evaluation

Real estate and capital investment analysis

Statistics

Taxes

In addition, an AICPA catalog of programs dealing with federal tax calculations is available on General Electric Company's time-sharing network [3]. Brief program descriptions are available elsewhere [2,3].

The General Electric Company, a leader in the time-sharing field, offers a relatively extensive range of programs and packages. The following list is representative of the services currently offered.

Accounting and Auditing

BUDGET$: Designed to reduce clerical load associated with expense budget preparation

CKREC$: Check reconciliation program; master data-base file maintenance of outstanding checks with file updating (new checks) and purging (returned checks)

GEPAY$: Maintains payroll, labor distribution, checks, payment summary, 941 and W-2 reports

GLAS: General Ledger Accounting System package; contains several programs and can be tailored to suit individual needs; provides a day-to-day accounting system

DEPREC: Calculates depreciation schedules and prints depreciation by months under four methods—straight line, double declining balance, sum of the years digits, and 150% declining balance

APSAM$: Appraises audit results by calculating confidence limits on cost questioned

RANUMS$: Generates random numbers within specified range

RASEQ$: Generates sets of random numbers

SAMSI$: Calculates necessary size of an audit sample to satisfy specified confidence limits on cost questioned

STATSYST: Statistical Analysis System package; provides statistical analyses including chi-square, regression, exponential smoothing, and moving averages without the necessity of user programming

In addition, the BMD Statistical Program package is available, as are specialized forecasting programs; e.g.,

DEPR: Provides a probability estimate of total revenue

Financial and Investment Analysis

ANNUIT: Calculates annuity payments/withdrawals and prints annuity tables

MORTGE: Calculates mortgage variables and prints monthly or annual tables

TRUINT: Calculates nominal, effective true interest rates for installment loans

TRUTH$: Calculates interest rate as required by the truth-in-lending act

PROFM$: Provides pro forma income statements and balance sheets for up to five future periods

FAPP: Financial Analysis of Projected Performance for up to 100 different projects; works as a data-base retrieval system and permits selective interrogation of financial status for the projects

CRNC: A 36-country currency exchange data base; display of new daily rates, time series of specific rate types

FAL II: Financial Analysis Language; produces financial reports and analyses to user specifications; pro forma financial reports, financial forecasting and budgeting, cash flow and capital structure analysis, tax analysis, capital budgeting, and investment analysis are available

CEIAP$: Capital Equipment Investment Analysis program; provides a comparison of one or more production methods. A probabilistic cash flow analysis is utilized

MERGA$: Analyzes alternative merger proposals

NICGO$: Calculates present/future stock value given abrupt changes in growth rate

PVROR$: Calculates present value, rate of return on investment

RETRN$: Calculates investor's rate of return on a stock for all possible past holding periods

SAVING: Calculates accumulated sum at the end of a desired time period at compound interest and with or without additions of capital

In addition to the software described above, there are many programs from contributing practitioners that have been placed in a General Electric Network Software Services catalog (for a description of this arrangement see [4]):

Financial Analysis and Planning Systems:

FCS: Financial and Corporate Planning System; available on Comshare's time sharing network; an extensive set of planning routines are provided; attractive features are sensitivity analysis provisions and the English-like commands, which make it easy to use

APL-PLUS/FPS: Financial Planning System, from Scientific Time Sharing Corp., Bethesda, Maryland; uses English-like commands and provides for sensitivity analysis; library of programs available.

AUTOTAB II: From Capex Corp., Phoenix, Arizona; language requires no programming language expertise; library of routines; has FORTRAN/COBOL interface.

B. Minicomputers and Small Computers

There has been significant product expansion in this area in recent years and there are currently many systems on the market (Chapter 5, Section IV, C). A selection that is relatively well known in accounting will be discussed here.

1. IBM 5100

Software available includes a client accounting and time management installed user program. The package is designed to run on an IBM 5100 Model B02 (32K processing unit) with an IBM 5103 printer attached. The 5100 is a desk-top computer, very compact, with a magnetic cassette reader incorporated in the main unit. The programs are written in BASIC and address the client write-up and internal time reporting requirements of the small certified public accounting firm.

A client's chart of accounts can be constructed by selecting from the standard chart of accounts provided and adding any unique accounts particular to the client if necessary. The entries for a client's accounting period are keyed into a transaction file and the data is edited, sorted, and processed to prepare the general ledger. Subsequently the balance sheet, income statement, and their supporting schedules may be produced. A statement of changes in financial position may be printed after preparation of the appropriate worksheet. A schedule of changes in working capital may also be produced.

The time management system aids the accountant in resource management and the preparation of client bills. Time reported by the accountant's staff can be keyed into the system daily or weekly, as can other client costs; the system can then generate a billing worksheet.

2. Litton 1241 and 1251

Litton provides basic accounting software packages that include the following: general ledger, financial statements, depreciation schedules, payroll processing, individual tax returns, invoicing and aging accounts receivables, accounts payable, job costing, and estimating. The software is designed to allow minor editing by the user. The hardware uses paper tape.

3. Burroughs L-2000

This is programmed in COBOL and allows the writing of user's own programs. In addition, software packages are available from Burroughs and several independent firms. R. J. Software Systems markets a client write-up package and a tax package, and Professional Software Consultants offers a package for the processing of financial statements and of 941 and W-2 statements.

The Burroughs L-2000, like the Litton 1241 and 1251, uses a paper tape reader. Paper tapes, being fragile, require careful handling and the equipment can be cumbersome. However, Burroughs also produces the L-8000, which uses either paper tape or a magnetic tape cassette.

4. Nixdorf 820/110

This hardware is modular, which provides for ease in modification or expansion; it also uses magnetic tape cassettes, which are more convenient and require less storage space than paper tape or punched cards. However, if additional peripheral devices, such as card or paper tape readers are desired, they can be added easily because of the modular nature of the system.

The software offered is labeled CLASS, the Client Accounting Service System. It has three modules: accounting, internal costing, and post-payroll. The accounting module provides the detailed listing of journals, the trial balance, the financial statements, updates the general ledger, and so on. The internal costing module helps measure the accountant's staff productivity and client profitability. Client billings are prepared automatically. The post-payroll module prepares the usual 941 and W-2 statements, as well as a payroll journal.

5. IBM System/3

There are several model configurations for this system, the variables being peripherals and size of cpu. For the packages presented here, the cpu requirement is in the range 8–64K of storage.

6. Online General Ledger and Financial Accounting System-II

This package is designed to perform general ledger accounting and furnish information for evaluating the performance of a business and requires a 64K processor. The programs are written in RPGII. Batch programs interface with on-line programs to produce financial statements; the on-line programs execute under control of the IBM Communication Control Program (CCP). Users can maintain books for many companies or clients, and financial reports for any company may be produced by selecting only that company's records. The package performs the following functions: transactions entry and edit, transaction register, fixed assets transaction register, trial balance, budget and forecast report, profit and loss statement, balance sheet, source and application of funds, statement of changes in financial position, summary of financial ratios, and year-to-date general ledger.

7. Fixed Asset Accounting and Control System

This system allows any combination of six depreciation methods:

Straight line

Double declining balance

Double declining with automatic switch over to straight line when it exceeds double declining

Sum of the years digits

Sum of the years digits with depreciation for 6 months in the first year

Sum of the years digits with depreciation for 12 months in the first year

The number of months to be depreciated may be set at 1, 3, 6, or 12. An asset master file is created and maintained on disk. Random update and retrieval is permissible, as well as transfers and additions. The program provides depreciation reports giving total accumulated, current year, and current period depreciation. Projections for the next 12 months are also available.

8. Public Budgeting and Financial Reporting System

This package is designed for cities, counties, and state departments or divisions. It consists of five modules: detail transaction processing, check writing, final reporting, budget preparation, and file maintenance.

9. Mini-Plan Financial Modeling Program

This program uses FORTRAN IV and comes with a library of subroutines that reduce the effort required to specify the logic of the application. In addition, a create program (to load and maintain files) and a print program (to generate a variety of reports) are supplied. The program has applications in short-range budgeting, cash planning, salary planning, and income and expense analysis.

10. Business Analysis/BASIC

This is a comprehensive set of 30 interactive analytical and data generation/maintenance routines written in BASIC for the System/3. The interactive features include instructional messages, flexible control of calculations, error checking, choice of methods for each routine, and data editing. The routines provided are basic to analysis activities and are designed to aid the financial decision-making process. The analytical routines may be used by themselves or together with the data generation routines to develop, output, and edit required financial reports. Brief descriptions of the routines are as follows:

Data Handling

$ID Identify: Index to name identification for each routine and file

$DF File: Creates and maintains data files

$DY Display: Prints data files

$BS Basort: Sorts data from a data file on a column

$SP Specify: Selects and sequences data from data file

$FM Format: Creates header information for any required report

$BD Build: Creates data file for report data

$TO Total: Creates data file for report data by adding or subtracting existing report data files

$CG Change: Used to maintain report data files

$WT Write: Used to print out the entire report

$EX Express: Faster mode of creating report data files for the experienced user

$VU Review: Allows user to determine the method of calculation used to create or change an existing report data file

Financial Analysis

$RI Return on investment: Allows choice of four methods to determine return on investment opportunities

$DC Discounted cash flow: Calculates the present or future value of an investment project

$ML Multiple loan analysis: Analyzes various loan alternatives and ranks them by selected criterion from the six available

$SL Single loan analysis: Analyzes components of a single loan; eight options are available to the user

$LS Lease vs purchase: Examines feasibility of leasing versus outright purchase

$MK Make vs buy: Outputs cash flows per period and present values of each alternative

$CT Break-even under certainty: Calculates break-even point for a deterministic situation

$UC Break-even under uncertainty: Calculates break-even with probabilistic data that the user provides

$DP Depreciation: Computes depreciation schedule and accumulated depreciation under straight line, declining balance (user-specified rate), sum of the years digits, and equipment units, the method being selected by the user

Time Series and Graphic Analysis

$GR Growth rate: Computes projections given simple or compound growth, the time period, the rate, and the initial value

$MA Moving average: Provides data-smoothing capability to reduce random fluctuation effects in a time series

$SA Seasonal analysis: Provides forecast lines with and without trend and seasonal indices, and computes mean absolute deviations to establish the best forecast line

$CA Cyclical analysis: Tests time series for cyclical patterns

$AC Autocovariances/autocorrelations: Computed for a given time series over various time periods

$LD Crosscovariance/crosscorrelation: Computes covariances and correlations between two sets of time series data

$SM Exponential smoothing: Computes a smoothed series of values given time series observations and a smoothing constant

$RG Simple regression: Computes the coefficients used to predict a variable in terms of an independent variable

$GM Histogram: Presents a bar graph for a row of the data file or for selected variables entered directly

11. Hewlett Packard 2000

An extensive contributed library of interactive programs for accounting, auditing, and financial analysis is available. In addition there are programs for data handling, statistics, management science, and forecasting. The whole library of contributed programs runs to several volumes; thus only a selection of the former will be presented here.

A total accounting system package, prepared by Datapoint Corporation, includes packages of program sets for accounts receivable, accounts payable, inventory control, payroll, and so on. Brief descriptions of programs from other contributors follow.

GSTKUL 36545 Stock valuation: Calculates PV of a stock based on alternative assumptions about growth rates for dividends and earnings, the terminal *P/E* ratio, and the relevant discount rate

LESSEE 36091 Lease analysis (Lessee): Compares leasing vs purchasing; sensitivity analysis available

LEASIN 36194 Lease income: Calculates lease income given number of units leased *U*, sales price *S*, at least rate *R*, for period *L*; sums income by year over *Y* years of operation

LOAN 36226 Loan amortization: Amortizes a loan on a monthly basis

IN/OUT 36088 Input/output analysis on economic flows: Divides a hypothetical economy into a certain number of industries using past period data; analyzes interindustry flows of goods and services over a period of time and predicts future flows under different conditions of consumer demand

MORGAG 36094 Mortgage analysis: Given three of the following four parameters, finds the remaining one—rate charged, life, amount borrowed, monthly payment

SAVING 36708 Compound interest: Computes the accumulated amount after *n* years at a given interest rate

STKINC 36096 Stock merger incentive program: Prints a table for stock incentive estimates, including prospective prices and gains, for the consolidated earnings of two companies considering merger

STKRTN 35098 Stock returns report: Computes a matrix of returns for an investment in a stock

STKVAL 36100 Stock value and evaluation report: Calculates a stock's value as determined by growth rate

TRUINT 36101 True annual interest rate analysis: Calculates true annual interest rate charged on an installment loan

GRISKA 36543 Risk analysis in capital investment: Compares alternative investments under uncertainty; provides a probability distribution of rates of return from the investments to allow the manager greater information for decision making

ANNUIT 36074 Annuity analysis: Determines payment or withdrawal annuities

BNDPRC 36076 Bond price analysis: Computes price *x* accrued interest for bond, given its coupon, redemption price, yield, and maturity

BNDSWH 36077 Bond switch analysis: Calculates effect of a bond switch, provides sensitivity analysis on various input

BNDYLD 36078 Bond yield analysis: Computes after-tax yield, given coupon, redemption price, maturity life, price, and tax rates on interest and capital gains

CAPDCF 36825 Capital investment analysis (DCF): Calculates rates of return using discounted cash flow method; user has choice of depreciation method; after-tax calculations

CAPINV 36080 Capital investment analysis: Provides a listing of gross cash flow, annual depreciation, annual tax, net cash flow, and discounted cash flow for a long term capital investment

DEPCOM 36082 Depreciation method comparison: Provides monthly depreciation by four methods: straight line, DDB, SYD, and 150% declining balance

EQUITY 36083 Cost of equity capital: Computes cost of equity capital by projecting dividends and share prices based on given growth rates and equating the PV of the stream to current share price

EXDRSK 36084 Extended risk analysis: Provides average cash flows per future period, expected pay back period, expected rate of return, and the probability of various rates of return

FINFLO 36711   Calculates present value: Assumes cash flows at end of period; calculates PV for given cost of capital

GCHLIN 36503   Rating investment funds: Fits a regression line to data on the quarterly rates of return for two entities; possibilities include mutual funds, individual stocks, indices of stock returns and portfolios; data files on mutual funds, dividends, etc. available for use with program

GDAPI 36507   Abnormal performance index: Computes this index based on price changes of stocks for which similar events have taken place

GDPA 36508   Efficient "corner" portfolios: Uses Markowitz's critical line algorithm to generate this set from a set of up to 100 securities

GIRRPV 36513   Investment return (cash flow): Simple program for calculating IRR and/or PV for sets of cash inflows and outflows

GKASSF 36514   Warrant price calculation: Calculates "normal" warrant price and "normal" change in warrant price per dollar change in price of associated stock

GKCOST 36515   P/E ratio Calculation: Calculates theoretical *P/E* ratio for a firm

GNMRVB 36530   Securities portfolio analysis and determination: Traces relationship between minimum nonmarket risk and market sensitivity for stock portfolios

GTHOR 36553   Securities EPS growth: Computes the number of years of constant growth in EPS required to justify current stock price

LOAN 36226   Loan amortization: Shows interest accumulated, payments to the principal, total paid, and remaining balance on a monthly basis

MARKOW 36092   Securities portfolio using Markowitz model: Computes the efficient securities portfolios according to the full covariance matrix Markowitz model

MCOST 36709: Compares and evaluates up to 1000 mortgage payment plans simultaneously

MKBUY 36093   Make–buy decision analysis: Prints a cash flow summary for each method for each year involved

III. Large System

One well-known large system will be covered to demonstrate the extra capability: the IBM 370. In addition to total accounting systems, the power of a large system makes feasible the use of comprehensive systems for extensive financial planning and analysis. The basic accounting software, as in the case of many of the smaller systems, is available from the manufacturer and from many software firms; it performs essentially the functions covered previously. Generalized audit packages are also available and, with the rapidly increasing capacity of small computers, some of these programs can be used on them.

Audit Packages

ASK 360: Developed jointly by Whinney Murray and Co. and Thomas McLintock and Co. of London, England; requires S/360-370 with at least a 32K core; the package is written in COBOL and assembler language

AUDASSIST: Alexander Grant & Co., Chicago; requires S/360-370 with a minimum core of 32K and is written in basic assembler language

AUDEX: Arthur Anderson & Co., Chicago; requires S/360-370 with at least a 32K core and is written in basic assembler language

AUDITAPE: Haskins & Sells, New York; requires S/360-370 with a 32K core; written in various languages and is amenable to use on smaller computers

AUDITPAK: Lybrand, New York; written in COBOL and thus requires a computer with a COBOL compiler

AUDITRONIC-16: Ernst & Ernst, Cleveland, Ohio; requires S/360-370 with a 32K core; written in basic assembler language and COBOL

AUDIT-THRU: Computer Resources, Inc., Wilton, Connecticut, written in COBOL and thus requires a computer with a COBOL compiler

AYAMS: Arthur Young and Co., New York; requires S/360-370 with at least 65K of core; written in basic assembler language and COBOL

CARS: Computer Audit Systems, Inc., East Orange, New Jersey; written in COBOL and requires a computer with a COBOL compiler

COMPUTER FILE ANALYZER: Price Waterhouse & Co., New York; requires S/360-370 with at least a 48K core; written in basic assembler language and COBOL

EDP-AUDITOR: Cullinane Corp., Boston; requires S/360-370 with a core of at least 53K; written in basic assembler language and COBOL

GRS: Program Products, Inc., Nanuet, New York; requires a computer with at least 64K of core and a COBOL compiler

MARK IV: Informatics, Inc., Canoga Park, California; requires S/360-370 with at least 65K of core; written in assembler language

MIRACL: Republic Software Products, East Orange, New Jersey; requires S/360-370 with at least 32K of core; written in assembler language and COBOL

SCORE: Programming Methods, Inc., New York; requires S/360-370 with at least 32K of core; written in COBOL and amenable to small computer use

STRATA: Touche Ross & Co., New York; requires S/360-370 with at least 32K of core; written in basic assembler language and COBOL

S/2170: Peat, Marwick, Mitchell & Co., New York; requires S/360-370 with at least 48K of core; written in COBOL and assembler language

Financial Analysis and Planning

PLANCODE Planning control and decision evaluation system: IBM; available in standard and interactive versions, this is a business management application program; provides a large library of financial analysis and statistical routines, and also has the ability to interface with other application programs written in FORTRAN, PL/1, COBOL, or assembler language

PSG II Planning Systems Generator: IBM; written in FORTRAN; a library of financial planning routines is provided; program assists in the generation of financial reports and in the analysis of a variety of planning problems; includes cash management, strategic planning, budgeting, marketing projections, balance sheet projections, etc.

BUDPLAN: IBM; the logic is similar to PSG II and the same application areas are addressed but it is written in PL/1

APL/FPS Financial planning system: IBM; conversational program; about 50 routines available; source language is APL but the user does not have to know this language

EPLAN Econometric planning language: IBM; written in APL and primarily for econometric use; emphasis is on time series forecasting

ABC Financial planning and forecasting: McCormack & Dodge Corp., Newton, Massachusetts; set of financial routines that are relatively easy to use

Capital investment & cost alternatives: MAN/VEST Consultants, Inc.; budgeting, cost analysis, forecasting, investment evaluation, etc.

ASC Sales Forecasting System: American Software & Computer Co.; time series forecasting; inventory control applications

BBL Basic business language: Sobco Corp., Wakefield, Massachusetts; written in TSO BASIC; financial routines, risk analysis, etc.

FPS: Rio Tinto North American Services, Canada; written in FORTRAN; financial routines, consolidation of reports

FIPAC: Multiple Access Computer Group, Canada; financial routines, planning, budgeting, and trend projections

FP/70: Bonner & Moore Software Systems, Houston, Texas; for reports and cash flow routines

SIMPLAN: Social Systems, Inc., Durham, North Carolina; interactive PL/1 package for financial planning and analysis

ADSIM: ADS Computing, Wellesley, Massachusetts; TSO FORTRAN; package has financial routines, reports, projections, consolidation, risk analysis

FORESIGHT: Foresight Systems, Inc., Los Angeles, California; FORTRAN; package has financial routines, reports, forecasting

REFERENCES

1. Cerullo, M. J., Untapped Computer Service Bureau Potential, *Journal of Accountancy* (December 1974).
2. AICPA Timesharing Program Library Available, *Journal of Accountancy*, pp. 95–96 (December 1975).
3. AICPA Timesharing Tax Catalog Available, *Journal of Accountancy*, pp. 97–100 (August 1975).
4. *Journal of Accountancy*, pp. 96–97 (October 1974).

Appendix 2
Abstracts of Recommended Articles* on Management Science Tools and Techniques

Back, Harry B., A Comparison of Operations Research and Management Science, *Interfaces* 4 (2), 42–52 (February 1974). This study explored similarities and differences between the journals *Operations Research (OR)* and *Management Science (MS)* by analyzing their citation patterns. The analysis suggests that *OR* and *MS* are very similar in the functions they perform. *OR* and *MS* have historically been devoted to closely related subject areas and appear to be currently carrying overlapping articles. The frequency at which topics not previously treated are introduced seems to be the same for both journals. The topics each journal covers are changing at approximately the same rate. The papers published in each journal are evidently of comparable significance. Despite these similarities, the analysis indicates that there are distinct differences in the roles played by *OR*: *MS* theory and *MS* application. Each publication seems to concentrate on a different set of subspecialties. In *MS*, topics not previously covered are brought in most often through the application series. *OR* is typically a source of information for both series of *MS* charts.

Blumstein, Alfred, Management Science to Aid the Manager—An Example from the Criminal Justice System, *Sloan Management Review* 15 (1), 35–48 (Fall 1973). After reviewing the literature on management science, one could justly describe the relationship between the practitioners of that art and the practicing manager as one of mutual disdain and distrust. All too often the dialogue between them has disintegrated into a rubble complete with parametric programming and regression analyses. Unable either to understand or to speak the jargon of management science, many managers have constructed a wall of diffidence between themselves and the management science profession. In this article, the author explores this problem within the context of the

* These articles were chosen as a result of search through data base 1971–July 1979. INFORM data base published by Data Courier, Inc., 620 South Fifth Street, Louisville, KY 40202. Telephone: (502) 582–4111.

criminal justice system. A computer model, which he specifically designed to have maximum appeal and usefulness for managers in the CJS, is described. The author is pushing even beyond the limits of detente. He seeks an active peace between managers and management science.

Bodily, Samuel E., Police Sector Design Incorporating Preferences of Interest Groups for Equality and Efficiency, *Management Science* 24 (12), 1301–1313 (August 1978). A decision model is proposed and illustrated by example for a resource allocation problem in urban management. It deals with the design of service areas for police mobile units. Estimates of the performance of alternative designs are provided by existing analytic models of spatially distributed emergency service systems. Making use of multiattribute utility theory, alternatives are evaluated according to the preferences of efficiency and equality of service of three interest groups: (1) citizens, (2) police, and (3) administrators. Meaningful measures of equality are addressed and developed, and an algorithm is created for generation of improved sector designs. Figures. Equations. References.

Bonini, Charles P., Computers, Modeling, and Management Education, *California Management Review* 21 (2), 47–55 (Winter 1978). The focus on computers in organizations has shifted from their use as merely a data-processing tool to a new place as part of a management information system (MIS) designed to support management decisions. At the same time, the operations research/management science (OR/MS) profession has expanded into the area of strategic management decisions and away from merely routine and structured decisions. The reasons for failure of computer models for top management are (1) lack of communication, (2) lack of management incentive, (3) difference in cognitive styles, (4) perceived threat of OR/MS/MIS, (5) state of the art, and (6) role of the analyst. In order for managers to successfully use computer modeling, they must (1) know OR/MS and MIS technology involved, (2) know how the manager will use the model, (3) have a sense of need for the model, (4) have a sense of proprietorship over the model, and (5) be able to deal with analysts. References.

Bruce, James W. Jr., Management Reporting System—A New Marriage Between Management and Financial Data Through Management Science, *Interfaces* (*TIMS*) 6 (1), 54–63 (November 1975, Part 2). This study reports a significant combination of a linear programming model of the LIBERTY, a communication system which visually depicts the model's output by electronically produced graphics, and a planning concept which has produced a new, more sensitive approach to asset/liability management of a banking organization in an increasingly volatile financial environment. During the past decade, banks generally have increased leverage and reduced liquidity, thereby making them increasingly vulnerable to interest-rate volatility. The management reporting system (MRS) described is a practical and unique approach in marrying the various sophisticated disciplines necessary for effective management given such constraints. A more realistic and sensitive approach to liquidity management developed at liberty is explored.

Carbone, Robert, and Longini, Richard L., A Feedback Model for Automated Real Estate Assessment, *Management Science* 24 (3), 241–248 (November 1977). Pressing changes are needed in the administration of real estate taxation that will not only ensure that all properties are assessed accurately and equitably, but that will enable taxpayers to perceive that they are being treated fairly. This study examines what properties an automated mass appraisal system should possess in order to meet the requirements of efficacy, equity, and public acceptability. A new automated system which utilizes feedback control and pattern recognition concepts is presented. The results of an empirical study using Pittsburgh data supports the feasibility of the proposed system. The basic goal behind automation is to provide equitable assessment roles. In the absence of observable fair market values, equity becomes a subjective judgment. Thus, a viable system must satisfy certain subjective public accountability requirements. Equations. Tables. References.

Chaiken, Jan M., Transfer of Emergency Service Deployment Models to Operating Agencies, *Management Science* 24 (7), 719–731 (March 1978). Six deployment models for emergency service agencies were developed, field-tested, and documented during a two-year period ending in October 1975. During the following 18 months, records were kept of the extent to which the models were acquired by operating agencies and actually used for making deployment changes. The number of acquisitions ranged from zero for one model to 39 for another. Over half of those who acquired these models actually put them to use and, except for one model, nearly all users made operational changes based on the output. Differences among the models illuminate the implementation process. Models are evaluated in such areas as data requirements, programming language, perception of impact, role of the advocate, and practitioner-to-practitioner transfer. Tables. References.

Debanne, J. G., Management Science in Energy Policy—A Case History and Success Story, *Interfaces* 5 (2), 1–21 (February 1975). The successful integration of management science in national policy formulation and decision making is well illustrated by the 1967 decision on the pricing of Canadian natural gas exports to the Pacific Northwest. The opportunity cost of exported gas was determined using dynamic programming and showed that the minimum cost alternative was much higher than the in-line price of Canadian gas exports imposed by the FPC. System simulations confirmed this finding and led to rejection of gas exports on FPC's terms and substitution of opportunity cost pricing to in-line pricing. The management science contribution was to show through optimal pipeline expansion studies that the alternative cost of U.S. sources of gas supply was higher. Thus rejection of the application would force abandonment of FPC's policy and lift the ceiling on gas export prices. Tables. Appendices.

Debanne, Joseph G., and Lavier, Jean-Noël, Management Science in the Public Sector—The Estevan Case, *Interfaces (TIMS)* 9 (2), 66–77 (February 1979, Part 2). The Canadian Department of Transport (DOT) commissioned a consultant's study that recommended the oldest of five ships then in service, The Estevan, be replaced, as well as suggesting that a sixth ship might be needed to meet the duty requirements of the Coast Guard. The Liberal government announced plans to build the new ship at the Vancouver shipyards during the 1978 election. The Deputy Minister of DOT commissioned an optimal investment study from the National Energy Board's operation research (OR) branch since it did not have time for an in-house OR group. The study was to determine which combination of features was to be a part of the design of the new ship, but the research study led to the conclusion that the Estevan should not be replaced and that only three ships were necessary to perform all of the Coast Guard's functions. The decision to build a new ship was, therefore, set aside. When the plan was implemented, it was credited with saving the cost of a vessel estimated at $6 million and with a gain in subsequent savings and investments many times that amount. Figures. Tables. Equations.

DeBrabander, Bert, and Edstrom, Anders, Successful Information System Development Projects, *Management Science* 24 (2), 191–199 (October 1977). It is generally agreed among researchers and practitioners that user involvement is the key to the success of computer-based information systems. This presentation treats user involvement within the framework of a dyadic (two-party) communication relationship between user and specialist. Previous work in the area is enlarged in three respects: (1) the theoretical framework is enriched by specifying what is considered to be important contingency factors for the interaction between user and specialist; (2) a formal theory for predicting the success of information system development projects is developed; (3) an important managerial tool to establish effective user–specialist communication is suggested. The basic assumption underlying the theory presented is that user and specialist bring different conceptual frameworks to the situation. Charts. References.

De Nisi, Angelo S., and Mitchell, Jimmy, L., An Analysis of Peer Ratings as Predictors and Criterion Measures and A Proposed New Application, *Academy of Management Review* 3 (2), 369–374 (April 1978). This study contends that advocates of peer ratings have been willing to ignore, or at least play down, the problems that have been shown to exist regarding the use of peer ratings. These problems should cast doubt on the utility of peer ratings as predictors of performance in all settings. An alternative application of peer ratings is presented which has the potential of minimizing the effects of some of the problems while still allowing benefits to be taken of some of their advantages. This new application of peer ratings is a source of feedback to the worker, who could thus obtain useful information about the way peers view his or her performance in terms to which he or she could probably relate well. It appears that more research is needed on the subject. References.

Dokmeci, Vedia F., A Quantitative Model to Plan Regional Health Facility Systems, *Management Science* 24 (4), 411–419 (December 1977). This presentation offers a quantitative planning model to determine the optimal characteristics (number, size, and location) of a regional health facility system. The system consists of a medical center, intermediate and local hospitals, and health centers. The quantitative model is based on the minimization of the total cost (sum of the transportation and facility costs) to the society. The optimal characteristics of the system are obtained using an heuristic method which includes both interactions of sublevel hospitals and environmental conditions as well. A numerical example is presented to illustrate the computational procedure. The quantitative model is presented for a region with no existing health facility system. If some health facility systems do exist, these may be evaluated initially. Equations. Tables. Figures. References.

Dyer, James S., A Time-Sharing Computer Program for the Solution of the Multiple Criteria Problem, *Management Science* 19 (12), 1379–1383 (August 1973). This note presents a description of a time-sharing computer program written to implement a human–machine interactive algorithm for the solution of the multiple-criteria problem. The interactive algorithm was suggested in a paper by Geoffrion, "Vector Maximal Decomposition Programming" (Working Paper No. 164, Western Management Science Institute, University of California, Los Angeles, September 1970). A unique feature of this program is the human–machine dialogue which obtains information from the decision maker through a series of simple, ordinal comparisons.

Ein-Dor, Phillip, and Segev, Eli, Strategic Planning for Management Information Systems, *Management Science* 24 (15), 1631–1641 (November 1978). Study is given to strategic planning for management information systems (MIS) with specific attention on (1) variables of the strategic plan which impact on the success or failure of MIS and (2) propositions relating states of the variables to system conditions. The variables addressed are (1) system development strategy, (2) the purpose of MIS, (3) the priority scheme, (4) functions assigned to the system, (5) goals, (6) definitions of requirements, and (7) documentation of the strategic plan. Two factors predominate in determinations of appropriateness of strategic plan for MIS: (1) explicitness and (2) situational fit. It is concluded that there exists no one optimal plan for MIS. Each organization must try to develop the strategy which best fits its unique situation. Tables. References.

Erlenkotter, Donald, Facility Location with Price-Sensitive Demands: Private, Public, and Quasi-Public. *Management Science* 24 (4), 378–386 (December 1977). A general model is formulated for an uncapacitated facility location problem in which demands are related to the prices established at the various locations. Pricing and location decisions are determined simultaneously, in contrast to traditional models that assume fixed demands and prices. It is shown that a transformation of the general model is equivalent to the fixed-demand location model of Efroymson and Ray. Specifying either a private sector objective of maximizing profits or a public sector one of maximizing net social benefits provides a particular case of the general model. A third plausible objective is the "quasi-public" one of social benefits with the constraint of suffcient revenues to cover costs. In this instance, a Lagrangian relaxation is utilized. Details of the solution approach are given for quadratic revenue functions, and an example illustrates the procedure. Tables. Equations. References.

Fuller, Jack A., and Atherton, Roger M., Fitting in the Management Science Specialist, *Business Horizons* 22 (2), 14–17 (April 1979). Due to the large number of quantitative techniques and tools being made available to organizations, managers are having to depend on staff specialists to help them choose and understand the appropriate tools to use in situations and to help them in evaluating and interpreting the results. Managers often fail to use quantitative techniques due to (1) a lack of knowledge of quantitative techniques, (2) no training in the use of quantitative techniques, (3) the high cost involved in using the techniques, and (4) the difficulties involved in quantifying the data needed for the techniques. Staff specialists often create problems by threatening line managers with their expertise. Problems can arise if the specialist approaches problems from a narrow rather than a broad point of view. Management science specialists who rely on the same analytical tools and techniques for every situation also create problems for the organization. The solutions for better staff specialists–manager interaction include (1) more informal and formal education for both groups, (2) better communication, (3) development of mutual trust and respect, and (4) the building of an increased awareness of how both groups can contribute to organizational effectiveness and accomplishment by working together.

Gardner, Bert, Electrify Your Thinking, *Accountancy* 85 (976), 86–89 (December 1974). Computer modeling systems permit the building of mathematical models of a company by noncomputer personnel. Financial modeling is a very powerful technique. Due to their involvement in budgeting and control systems, accountants are often better placed than those of most other disciplines to have a wide understanding, and often unique knowledge, of how their organization functions. If accountants fail to master the new technique, with the opportunity it offers to harness this knowledge and help management in overall control and direction of their organizations, then the future status of the profession may suffer. Many others who are not so qualified are only too ready to present it as a new management science when it is, in fact, only electrification of the methods accountants have used for years. Tables.

Ginzberg, Michael J., Steps Toward More Effective Implementation of MS and MIS, *Interfaces* 8 (3), 57–63 (May 1978). Many sophisticated management science models and management information systems have been developed which have not worked well in the real world. Early scientific approaches to implementation, the normative and factor approaches, were unsuccessful. A more successful approach is the dynamic process model which analyzes the best opportunities to influence designer or user behavior. It is based on the planned change approach, which is a systematic approach to unfreezing existing beliefs, changing them, and refreezing the correct beliefs. Empirical studies show that situations conducive to planned theory change have a higher successful implementation record. To be successful, designers should take special care to ensure that they address the needs of the user. Users should understand that a successful project will require changes in procedures and attitudes. References.

Glover, F., Hultz, J., Klingmand, et al., Generalized Networks: A Fundamental Computer-Based Planning Tool, *Management Science* 24 (12), 1209–1220 (August 1978). The recent emergence of generalized networks as a fundamental computer-based planning tool is documented, and the power is demonstrated of the associated modeling and solution techniques when used together to solve real-world problems. A nontechnical account is presented of how generalized networks are utilized to model a group of significant practical problems. In addition, a technical exposition is presented of the design and analysis of computer solution techniques for large-scale generalized network (GN) problems. These contain an investigation of GN solution strategies within the framework of specializations of the primal simplex method. Degeneracy rules are also identified. An efficient solution procedure is offered derived from an integrated system of start, pivot, and degeneracy rules. It is shown that the resulting computer code is at least 50 times more efficient on large problems than the LP system APEX-III. Table. Equations. Figures. References.

Graham, Robert J., Problem and Opportunity Identification in Management Science, *Interfaces (TIMS)* 6 (4), 79–82 (August 1976). One of the major problems in the development of OR/MS has been the development of problem identification techniques. This study of the use of computer models for problem realization describes the process of formulating a computer model designed to identify problems rather than to offer solutions. The principle problem attacked was that of a reactive attitude towards the future. That is, management seems to want the future to be pretty much like the immediate past. Management was not against change but it did resist change. The process and subsequent model results developed showed that results of current trends were disastrous over a ten-year period and that propagation of change was the only worthwhile strategy to consider.

Graham, Robert J., The First Step to Successful Implementation of Management Science, *Columbia Journal of World Business* 12 (3), 66–72 (Fall 1977). Management science models can prove much more useful if managers and management scientists first work together to define the problem and then build the appropriate model. The system of problems must first be defined, management must participate, problem sources must be identified, and all groups with a share in the organization's future must be analyzed. In identifying a system of problems, as many problems as possible should be pinpointed and their interrelationships studied. As many managers as possible should interact to isolate problems, rather than debate possible solutions. The combination of people and events leading to a problem situation should be identified to find the source of the problem. Groups affecting or affected by the larger system should be examined to determine the consequences of their actions on the system. A checklist integrating all of the important variables should be made for every problem defined. Diagram.

Greco, Richard J., MIS Planning—An Approach, *Data Management* 9 (10), 17 (October 1971). GAC Corporation has an organized, structured approach to MIS planning. There are three major data centers and a staff group reporting to the corporate Vice-President. Each center has development, operations, and technical support functions. The MIS staff is composed of specialists in business information systems, computer systems, data-base management, and management science. Business systems consultants interface subsidiary operating and staff departments, and regional and corporate staffs. Technical handles computer systems utilization and support. Data-base consultants are responsible for base design, management, security, and operations. Management science concerns itself with quantitative tools and techniques for decision making in the firm. MIS planning structure and operations are described.

Greenlaw, Paul S., Management Science and Personnel Management, *Personnel Journal* 52 (11), 946–953 (November 1973). The purpose of this article is to demonstrate that personnel management is no longer one of the weaker functions of management. A number of management science techniques have been developed for personnel. Models have been designed for such personnel functions as recruiting, selection, manpower planning, safety, wage and salary administration, and collective bargaining. Three studies used automated or cognitive process simulation in personality assessment and/or personnel selection. Linear programming and PERT have been applied to college recruiting. Five Markovian models and the logarithmic learning curve model have dealt with the area of manpower planning. The logarithmic model has been applied to safety and Simplex to work salary administration. There is one application of PERT, one simulation model, and four linear programming models.

Gupta, Jatinder N. D., Management Science Implementation—Experiences of a Practicing OR Manager, *Interfaces (TIMS)* 7 (3), 84–90 (May 1977). This presentation is a commentary on the experience of a practicing OR manager involved in the application of management science in a large public enterprise. These experiences demonstrate that implementation is not the aftermath of a scientific investigation; rather, it is an integral part of the problem-solving and system design process. Several practical difficulties are analyzed which are encountered in the OR/MS application. Then an implementation strategy is suggested which proposes reexamination of the optimality concept and the interdisciplinary teamwork approach to OR/MS implementation. Among the strategic steps suggested are the following: (1) Each decision should be analyzed in terms of the consequences on decision factors. (2) Cause and effect relationships must be established for decision factors influencing the thinking of a manager. References.

Gupta, Shiv K., A Language for Policy-Level Modelling, *Journal of Operational Research Society* 30 (4), 297–308 (April 1979). The efficacy of operational research/management science (OR/MS) modeling techniques has not been widely demonstrated as aids in policy/strategy formation, in spite of successful applications to operational-level problems. It is suggested that the language of traditional OR/MS model structures is too restrictive to deal with issues typically confronted in policy making. This study (1) discusses differences between operational-level decision-making processes and policy-strategy formulation processes, (2) describes and classifies alternative model structures, and (3) proposes a modeling language for addressing issues of policy. This language is presented in set-theoretic terms, the resulting structure implying that certain model classes accommodate policy-level management needs more effectively than others. It is further hypothesized that policy-level models must serve a role district from the prediction and optimization role typically addressed by operational-level models. Figures. Tables. References.

Higgins, J. C., and Finn, R., Managerial Attitudes Towards Computer Models for Planning and Control, *Long-Range Planning* 9 (6), 107–112 (December 1976). While it is common to cite lack of numeracy as one of the principal reasons for management disinterest in operational research models, there are other fundamental causes of this phenomenon. At the senior executive level, the nature of the role and the individual's style may be the dominant factors. Many senior managers seem to feel that the processing of numerical information is not a major feature of their role. A crucial aspect of the senior manager's job is an acute lack of time. This results in the analyses of strategic decisions being performed by subordinates and in senior managers not having sufficient time to gain the necessary appreciation of a particular model. The economic environment in which the company is operating appears to affect the use of management science techniques. The more unstable the environment, the more managers rely on intuitive judgment as opposed to scientific methods.

Hoffman, Gerald M., The Contribution of Management Science to Management Information, *Interfaces (TIMS)* 9 (1), 34–39 (November 1978). Management scientists (MS) and operations researchers (OR) should become deeply involved in the information systems activity, because MS can make major contributions to decisions that are totally outside such systems. Operations research is the application of mathematics to management problems. Management science is the application of all sciences to management problems. Management information systems (MIS) focus on managerial decisions. Their function is to capture data, process it into a form useful to the managers, and transmit it to them in a timely and useful way. MS is not a set of techniques, but a process of changing the management decision process. References.

Holstein, William K., and Berry, William L., The Labor Assignment Decision— An Application of Work Flow Structure Information, *Management Science* 18 (7), 390 (March 1972). The movement of workers among machines is a tactic often employed by job-shop managers to break bottlenecks and smooth the flow of work. Despite its common occurrence in industrial practice, the problem has only recently received attention in the management science literature. The purposes of this paper are (1) to discuss the nature of the labor assignment problem in job shops and (2) to suggest a procedure for making labor assignments at the time of actual production. The significant features of the specific rules suggested are that they use aggregate information on work flow patterns in the shop and that they make fewer labor transfers than other rules which have been suggested.

Konczal, Edward F., Documenting Large-Scale Telecommunications Computer Analyses, *Journal of Systems Management* 29 (6), 14–17 (June 1978). Corporations and governmental managers have begun to seek management science analytical models to assist in decision making. These models utilize linear programming, inventory models, and break-even analysis. More complex models use a combination of management science methods. The validation of these complex computer analyses uses no definitive, universal validation test. Validation models utilize either qualitative or quantitative methods. Qualitative procedures are concerned with investigating logical consistencies and worthiness. Components of these procedures include documentation packages, technical text, system flow charts, and computer code. Quantitative measurements are more objective, using statistical and sensitivity tests. Sensitivity tests should consider the nature of the testing procedure, the number of tests, and the components analyzed. Benchmark computer programs should be established as a basis for measurement. Charts. References.

Liberman, Varda, and Yechiali, Uri, On the Hotel Overbooking Problem—An

Inventory System with Stochastic Cancellations, *Management Science* 24 (11), 1117–1126 (July 1978). The problem addressed is finding the optimal overbooking strategy to be followed by a hotel in order to maximize net profit. It is demonstrated that the optimal strategy is a three-region policy. The three regions are presented. For each period, there exist upper and lower bounds and an intermediate point such that (1) if the overbooking level at the end of a period is greater than the upper bound, it should be decreased to that bound; (2) if the inventory level is below the lower bound, two cases may occur involving taking additional reservations to equal the lower bound and keeping to the intermediate point in respect to new requests; (3) if the inventory level is between the two bounds, two possibilities exist of not confirming new requests if it is above the intermediate point or confirming some new requests provided the new inventory level will not exceed the intermediate point. Equations. References.

Lorsch, Jay W., Making Behavioral Science More Useful, *Harvard Business Review* 57 (2), 171–180 (March/April 1979). Behavioral science theories have potential application to management. However, many of these ideas are only sparingly used. Occasionally, managers have applied concepts from psychology and have been disappointed with the results and, therefore, hesitant to try new theories. One of the problems in applying knowledge from the behavioral sciences has been the interpretation that the ideas are applicable to all situations. While most managers prefer the simpler, quicker prescriptions of universal theories and techniques, it is important to realize that each situation is unique and that universality may be inappropriate. The behavioral sciences can provide situational theories to help managers analyze, order, understand, and deal with the complex social and human issues they face on the job. Managers need to become critical consumers of behavioral theories and analyze and diagnose in order to develop skills and knowledge to use the available tools. Charts.

Lucas, William, An Overview of the Mathematical Theory of Games, *Management Science* 18 (5), 3 (January 1972, Part II). A cursory survey of the theory of games is presented. The basic models for games in extensive form, normal form, and characteristic function form are discussed, and some generalizations and extensions of these models are mentioned briefly. Some general remarks about the applicability of the theory are made in two cases, and a few specific applications are referred to in sections subsequent to the description of the particular models. Mention is made throughout this paper to the other game-theoretic articles in this issue in an attempt to place them somewhat within this outline of the subject.

McNally, James P., How Management Science Can Save You Money, *Price Waterhouse Review* 22 (2), 18–23 (April 1977). "Results-oriented" management science applications can prove simple, practical solutions rather than being too theoretical, as many managers believe them to be. One Price Waterhouse engagement saved a client millions of dollars yearly by solving its inventory management problems via management science techniques. The client had 16 branches across the U.S. which independently ordered inventory. A study revealed that the company could save money by formalizing its ordering rules and establishing a joint-purchasing system for the branches. The study was conducted in the following phases: First, to see if a mechanized inventory system should be used to support purchasing, new purchasing rules were tested in actual use. Then a computer-based system was established to support the manual system, and rules for joint purchasing were written and adopted. Chart.

Montanari, John R., Managerial Discretion: An Expanded Model of Organizational Choice, *Academy of Management Review* 3 (2), 231–241 (April 1978). Managers are continually seeking ways to deal with changing business conditions, and one method of adapting the organization to cope with complex and dynamic business situations is structural modification. In this study, a review of literature on structural determination is accompanied by a discussion of recent attempts to incorporate managerial discretion into the existing structural model. Lacunae in current conceptual formulations are highlighted. An expanded choice model is proposed which incorporates prior empirical research and contemporary theoretical developments into a conceptual paradigm of the strategic decision-making process. The model proposes organizational size, environmental conditions, and technology as co-acting influences on organizational and managerial factors. Figures. References.

Moyer, M. S., Management Science in Retailing, *Journal of Marketing* 36 (1), 3–9 (January 1972). Retailers who would like to employ operations research techniques should understand that successful applications are fewer in retailing than in other areas of marketing; but management will find some new opportunities to improve profits. Promising applications for improving decision making through OR are currently being proposed that pertain to most elements of the merchant's marketing mix. In areas of pricing, sales-force planning, credit granting, and store development, practical models are now available. The main impediment to management science aid may be the inability or incapacity of management to use the techniques.

Nackel, John G., Goldman, Jay, and Fairman, William L., A Group Decision Process for Resource Allocation in the Health Setting, *Management Science* 24 (12), 1259–1267 (August 1978). A decision process is described that allocates an available budget to health programs based on program effectiveness. Effectiveness is measured by comparing alternative programs at various funding levels in regard to the weighted objectives of the health care organization. Organizational objectives are weighted via a constant-sum paired comparisons scaling technique. The decision makers can maximize program effectiveness with an integer programming function within the given budgetary, resource, regulatory, and program structure constraints. The effectiveness evaluation–resource allocation process is validated by means of actual decision-making teams testing four processes for resource allocation. Tables. Equations. References.

Noble, Carl E., Why Wait for Crises to Reveal the Unusual Opportunities for Management Science Applications in your Organization? *Interfaces (TIMS)* 7 (4), 50–55 (August 1977). In this presentation, four case studies describe how the application of well-known statistical methods produced relevant and action-oriented information and control systems. In turn, these systems made large contributions to the profits of the firms involved. The factories which were operating in the case studies were apparently in healthy states until crises revealed the inadequacy of their information and control systems. Investigations following the crises revealed that (1) the efficiencies of the factories were well below their potentials, even though they appeared healthy, (2) management was making decisions without meaningful information and/ or without knowledge of large variations in testing and production processes, and (3) the information and control systems often lacked valid links with either suppliers or customers. The management scientist must not wait for crises to bring opportunities to the surface. References.

Paul, Robert J., The Retail Store as a Waiting Line Model, *Journal of Retailing* 48 (2), 3–15 (Summer 1972). The scheduling of personnel in service activities and variable tasks has historically fallen into the realm of subjective judgment where the art of management has overshadowed the science of management. The determination of a manpower schedule for variable service activities has been hampered by the absence of formal standards and delay costs and by the high cost of study and measurement. However, with some recent developments in work-measurement concepts combined with applications of management science techniques, it may be possible to develop objective schedules for the variable service-type activity, as well as for the repetitive manual task.

Pike, Dan, Management Theory: Its Application to the Job, *Supervisory Management* 23 (12), 26–30 (December 1978). Supervisors can benefit from studies of both management and behavioral science. Success in today's technically oriented business world is dependent upon an optimal mixture of the two. One of the building blocks of management science is Douglas McGregor's Theory X, Theory Y hypothesis. Maslow's theory of hierarchy of needs is also important to management science development. Theory X, Theory Y is based on the view that the manager has in regard to whether employees need constant supervision or whether they can work independently because they are creative and intelligent. The manager must determine how to best combine the qualities of behavioral science. In such an instance as delegation of a task, the manager must follow certain steps: assign the task and specify the parameters for it, then assign accountability and exact responsibility. Management science is modified by the actual work place. Management science must be applied to the work environment and behavioral science to management science in order to lead to effective methods of managing.

Render, Barry, and Stair, Ralph, Management Science and the Small Business, *Journal of Systems Management* 28 (3), 20–22 (March 1977). Management science (MS) is an approach to decision making in which the principles of the scientific method are utilized to find solutions to business problems. A manager's analysis process may take two forms: qualitative or quantitative. MS concentrates on quantitative data relating to a problem. Quantitative tools include network models, forecasting techniques, inventory models, simulation, mathematical programming, decision analysis, and queuing models. Recent years have seen an increase in the availability of MS techniques. The arrival of high-speed computers, capable of easily handling the sophisticated equations of MS, has paralleled the advance of MS as a management tool. Tables.

Sharon, Ed M., Decentralization of the Capital Budgeting Authority, *Management Science* 25 (1), 31–42 (January 1979). Delegation of authority for capital budgeting is of major interest to business management. A conceptually new first-cut decision aid is presented for selecting the appropriate degree of decision authority to be delegated to divisions in an hierarchical organization. The decision aid takes a subjective approach to capital budgeting from the perspective of the delegator. A model serves to capture the trade-offs between the cost of centralized (duplicated) review and the opportunity cost and risk of having (otherwise) "unacceptable" projects. Identification is made of two explicit limits to optimal decentralization to help headquarters (HQ) select the kinds of proposal that should be (1) approved and executed by divisions without HQ review, (2) submitted to HQ for centralized review, and (3) rejected by divisions without HQ review. The process is demonstrated by a numerical example. Tables. Equations. References.

Schneyman, Arthur H., Management Information Systems for Management Sciences, *Interfaces (TIMS)* 6 (3), 52–59 (May 1976). Management information systems for management science (MS) functions should deal with the management process and decision making as applied to internal and external resources of both the MS unit involved and the host organization. Management information systems should recognize the basic MS organization unit as consisting of data center, applications, and planning and control groups, all viewed as an MS performance center. It is vital to comprehend MS goals, and information systems should be structured in terms of net contribution to the host organization. They should measure the productivity, quality, reliability, and service of MS cost performance. Charts.

Schroeder, Roger G., *A Survey of Management Science in University Operations* (1973). This survey discusses applications and research of the management sciences in institutions of higher education. It includes four specific areas: (1) planning, programming, and budgeting systems (PPBS), (2) management information systems (MIS), (3) resource allocation models, and (4) mathematical models. In each of these areas, the major trends and literature are discussed along with numerous references. The survey also provides an integration and synthesis of the recent work from both published and unpublished reports. The authors conclude with four specific problem areas which are in need of further attention: (1) stability and suitability of various student flow projections, (2) investigation of decision-making processes and the information which should be used, (3) measurement of output, and (4) alternative approaches to improve planning methodology.

Shycon, Harvey N., Perspectives on MS Applications, *Interfaces* (*TIMS*) 4 (4), 39–42 (August 1974). The placing of value on the management science activity and the measurement of its contribution to the business/industrial organization is examined in this article as one of the most urgent considerations confronting management scientists and operations researchers today. Measurement of the value of management science is divided into three areas—the tangible or obvious, the semitangible, and the intangible or obscure—with the greatest contribution of management science believed to lie in the area of intangible contributions. It is suggested that the real measure of success is within our own organizations, and that the final test of the value of management science to the firm may be in how often decision-making management calls upon the management science group for assistance.

Shycon, Harvey N., Perspectives on MS Applications, *Interfaces (TIMS)* 6 (4), 27–30 (August 1976). An operating strategy for management science groups in a rising economy must be based on cautious optimism. There are three factors around which an operating strategy during an emerging economy should be built. These are project selection, communication, and the management science organizational structure. There is a fourth factor which helps prepare for greater gain in the future: education for management on the potential offered through management science, using this as a vehicle to obtain management involvement. No function within the organization is as well qualified as the management science group to view the firm as a whole and to help determine how best to move ahead as the economy recovers.

Sprague, Linda G., and Sprague, Christopher R., Management Science, *Interfaces (TIMS)* 7 (1), 57–62 (November 1976, Part 1). Academic-based management scientists can work on real or toy problems, and they can conduct real or toy research. It is hypothesized that the bulk of articles appearing in journals are, in fact, either toy research on real problems, or real research on toy problems. What is needed, obviously, is real research on real problems. Research in the field of management is in its infancy compared with other disciplines. Research in the field has evolved thus far along three dimensions: data exploitation, experimentation, and abstraction. Management scientists are doing good work today in their research efforts. Unfortunately, the bulk of their science never finds application, and their engineering is never brought to professional associations. Management scientists must preach what they practice and strive harder to bring their fascinating work out of the closet and into print. References.

Stradtherr, Mark A., and Rudd, Dale F., Resource Management in the Petrochemical Industry, *Management Science* 24 (7), 740–746 (March 1978). As petroleum and natural gas supplies dwindle, the petrochemical industry must look elsewhere for its raw materials. A linear programming model for the industry is described in which the objective is the efficient use of feedstocks. Four feedstock alternatives are analyzed with respect to changes in the optimal process technology resulting from changes in feedstock availability. The model solutions also indicate the net change in feedstock requirement for the industry per unit increase in demand for each product. This "feedstock efficiency index" draws attention to areas of incentive for the development of new process technology. These areas are different for different feedstock alternatives. This sort of analysis should provide government planners with information needed for long-range planning and resource management. Equations. Tables. References.

Vazsonyi, Andrew, Information Systems in Management Science–Decision Support Systems: The New Technology of Decision Making, *Interfaces (TIMS)* 9 (1), 72–77 (November 1978). Many managerial decisions require quick responses from decision support systems. The major limitation of operations research/management science (OR/MS) is the length of time needed to complete an OR/MS study. A solution to this problem lies in a new technology called decision support systems (DSS), a type of information system. This development is different from traditional computer-based approaches to problem solving. DSS relies on the decision maker's insights and judgment from problem formulation to evaluating the solutions presented. DSS deals with unstructured managerial problems. Design criteria for DSS involve (1) using CRTS for report generation, charts, graphs, etc., (2) support of intelligence, design, and choice alternatives, (3) data-based management systems that allow flexible, interactive access to data, and (4) human–machine interaction without programmers. This technology can successfully reduce the time and cost of the decision-making process. Table. References.

Watson, Hugh J., Stimulating Human Decision Making, *Journal of Systems Management* 24 (5), 24–27 (May 1973). One increasingly valuable use of the computer is to assist management in finding solutions to nonroutine problems. In order to accomplish this objective, management science techniques are being integrated with the computer to give management more and better information than ever before. For some applications management does not even need to know how the method of analysis is performed, but rather only how to structure the problem properly and supply the requisite data inputs to the computer. Another evolving use of the computer involves the simulation of human decision-making and problem-solving processes. These are computerized models that duplicate, or perhaps even improve upon, the decisions of human beings.

Weinberg, Charles B., Jointly Optimal Sales Commissions for Nonincome Maximizing Sales Forces, *Management Science* 24 (12), 1252–1258 (August 1978). It has been shown previously, with regard to a multiproduct sales force, that a commission structure based on equal fractions of each product's realized gross margin is jointly optimal if the goal of the sales force is to maximize expected income and the products are independent. The sales force will act jointly to optimize its own objective and maximize corporate earnings. It is demonstrated that an equal gross margin commission system is also jointly optimal when products are interdependent and when the sales force does not have a goal of income maximization but embraces other objectives such as minimizing time to reach a certain income goal or trading-off time against money. Not all salespeople are required to have the same objective. An approach is developed for income maximizers which require less stringent assumptions than previously employed. Equations. References.

Weinberg, Charles B., and Shachmut, Kenneth M., ARTS PLAN: A Model Based System for Use in Planning a Performing Arts Series, *Management Science* 24 (6), 654–664 (February 1978). This presentation addresses the application of marketing science to some of the problems involved in managing a series of performing art programs. Although a portion of the decisions are largely "artistic" and not applicable to formal analytical procedures, many others can be aided by model-based procedures. This study describes some of the analytical procedures that are being utilized by management of one performing arts series at Stanford University. The starting point for ARTS PLAN is a dummy regression variable analysis of historical data which leads to a prediction of attendance at future performances. The manager can override the prediction if he disagrees with the forecast. The forecast is then embedded in an interactive model which can be used in planning a season or in deciding which events to promote in a season that has already been planned. Tables. Equations.

Index